DATE DUE

JUL 2 4 1980		
DEC 0 8 1980		
MAR 3 0 1984		
MAR 1 6 1989		
FEB 1 1 1992		
FEB 1 8 1995		
JUL 1 8 1996		

Biochemistry and Physiology
of Plant Hormones

Thomas C. Moore

Biochemistry and Physiology of Plant Hormones

With 164 Figures

Springer-Verlag
New York   Heidelberg   Berlin

Thomas C. Moore
Professor of Botany
Department of Botany and Plant Pathology
Oregon State University
Corvallis, Oregon 97331
USA

Library of Congress Cataloging in Publication Data
Moore, Thomas C.

Biochemistry and physiology of plant hormones.

Includes bibliographies and index.
1. Plant hormones. I. Title.
QK731.M66   581.1'4   79-11492

Printed in the United States of America.

987654321

ISBN 0-387-90401-8 Springer-Verlag   New York
ISBN 3-540-90401-8 Springer-Verlag   Berlin   Heidelberg

Preface

Biochemistry and Physiology of Plant Hormones is intended primarily as
a textbook or major reference for a one-term intermediate-level or advanced course dealing with hormonal regulation of growth and development of seed plants for students majoring in biology, botany, and applied
botany fields such as agronomy, forestry, and horticulture. Additionally,
it should be useful to others who wish to become familiar with the topic in
relation to their principal student or professional interests in related
fields. It is assumed that readers will have a background in fundamental
biology, plant physiology, and biochemistry.

The dominant objective of *Biochemistry and Physiology of Plant Hormones* is to summarize, in a reasonably balanced and comprehensive
way, the current state of our fundamental knowledge regarding the major
kinds of hormones and the phytochrome pigment system. Written primarily for students rather than researchers, the book is purposely brief.
Biochemical aspects have been given priority intentionally, somewhat at
the expense of physiological considerations. There are extensive citations
of the literature—both old and recent—but, it is hoped, not so much docu-
mentation as to make the book difficult to read. The specific choices of
publications to cite and illustrations to present were made for different
reasons, often to illustrate historical development, sometimes to illustrate
ideas that later proved invalid, occasionally to exemplify conflicting hy-
potheses, and most often to illustrate the current state of our knowledge
about hormonal phenomena. The lists of references at the ends of the
chapters, containing some references which are cited and others that are
not, are not intended as comprehensive bibliographies of the most recent,
or even exclusively the most important, publications on each subject.
Each list is intended both to document the text and provide other examples of the extensive literature on each topic.

An explanation should be given for inclusion of the subject matter comprising Chapter 1, since it is acknowledged that many readers will regard Chapter 1 as quite elementary information with which they already are familiar. That is fully to be expected. But for those readers whose background may be deficient, as has been found to be true of a fair percentage of students, Chapter 1 will provide a reasonable overall introduction to and perspective about growth and development of whole plants throughout ontogeny and set the stage for consideration of hormonal regulation.

Books such as this invariably disappoint some readers, which is to say that they cannot be—perhaps should not even purport to be—all things that all readers might wish or expect. In my judgment, *Biochemistry and Physiology of Plant Hormones* most likely might disappoint some readers in each of two ways. First, the book does not contain as lengthy and integrated a discussion either of the physiological roles of the different kinds of hormones or of hormonal interactions as some readers will wish, although, of course, these topics definitely are covered. To the extent that this is true it is by design. For in my ten years of experience teaching a graduate course in hormonal regulation of plant growth and development, I personally have found that it is more effective to guide students from an information base such as this book provides to a more integrated understanding of regulation of growth and development than to undertake the converse approach. Another way this book might disappoint some prospective users is that it lacks detailed and comprehensive coverage of practical uses of synthetic plant growth regulators, except for synthetic auxins and auxin-type herbicides. Such information is largely beyond the scope of this small volume. Moreover, practical uses of plant growth regulators are covered in many specialized books in agronomy, forestry, and horticulture.

It seems to be a good time in some ways, and not so good a time in other ways, for a new book on the biochemistry and physiology of plant hormones. On the negative side, so far during the decade of the 1970s there seems to be a relative lull in the field as regards dramatic new developments—the "acid growth theory" and other important advancements notwithstanding—compared, let us say, to either of the previous two decades. In view of the relative scarcity of "big news," it could be argued that it is not a particularly good time. But, on the other hand, there is really good and highly significant research going on, and there is a steady output of important new knowledge. The literature—the state of the science—probably is in the best shape ever as far as unequivocal validation of facts and concepts is concerned. It is a time of separation of fact from fiction and devising new approaches to old problems, as well as asking new, important, exciting questions. For these reasons, it seems, therefore, timely for a new book to call attention to this healthy state of the science. In any case, it is an excellent time to be a student at any level of the fascinating subject discussed in *Biochemistry and Physiology of Plant Hormones*.

Acknowledgments

The real credit for *Biochemistry and Physiology of Plant Hormones* ultimately should go to the many Plant Physiologists whose research during the last half century disclosed the information comprising the book. While too numerous to mention individually, the names of many of these scientists are contained in the literature lists at the ends of the chapters. Certain specific contributions by particular authors, of course, are acknowledged also in the forms of citations in the text and notations in legends to figures and tables.

The actual writing and production of the book naturally has involved several forms of assistance by many persons, to all of whom I express my sincere gratitude. For directly supplying or assisting to make available certain illustrations, I thank Douglas O. Adams, James D. Anderson, Gerard W. M. Barendse, Michael L. Evans, Peter Hedden, Hans Kende, Anton Lang, A. Carl Leopold, Morris Lieberman, Bernard O. Phinney, Folke Skoog, Nobutaka Takahashi and Jan A. D. Zeevaart. Donald J. Armstrong and Ralph S. Quatrano are thanked for the advice and technical assistance that they provided regarding various topics. I thank Ellen Witt and Leona Nicholson for typing and clerical assistance, and E. Kay Fernald for photographic service. Mark Licker, Science Editor, and Judi Allen, Production Editor, at Springer-Verlag Inc., New York, and their staff were very helpful throughout the review and production processes. Finally, I wish to acknowledge the financial support provided by the National Science Foundation for those of my own investigations during the past fifteen years which are cited in the book.

Corvallis, Oregon   Thomas C. Moore
January, 1979

Contents

Chapter 3

Gibberellins 90

Chapter 4

Cytokinins 147

Chapter 1
Introduction

Fundamental Terms and Concepts

The term "development," as it applies to whole seed plants arising by sexual reproduction, denotes the gradual and progressive changes in size, structure, and function which collectively comprise the transformation of a zygote into a mature, reproductive plant. It is also a correct and common practice to speak of the development of particular organs from initials or primordia, and to refer to development of a whole plant from any single cell. Whatever the specific case, development is a gradual process that takes time to be fully realized, generally is accompanied by increases in size and weight, involves the appearance of new structures and functions and the loss of former ones, is characterized by temporal and spatial discontinuities and changes in rate, and eventually slows down or ceases when mature dimensions are reached.

Unfortunately, but perhaps not surprisingly, there is no rigorously standardized terminology applied to the phenomena of plant growth and development. Some physiologists, the author included, consider that there are three interrelated processes that together comprise development, namely, "growth," "cellular differentiation," and "morphogenesis." "Growth" is defined as an irreversible increase in size which is commonly, but not necessarily (e.g., the growth of an etiolated seedling), accompanied by an increase in dry weight and in the amount of protoplasm. Alternatively, it may be viewed as an increase in volume or in length of a plant or plant part. In any case, it must be emphasized that growth can occur only by an increase in volume of the individual cells. Some authors consider cell division as a separate process which accompanies growth in meristems, but a more generally held view is that growth includes cell division as well as cell enlargement.

"Cellular differentiation" is the transformation of apparently genetically identical cells of common derivation from a zygote or other single cell into diversified cells with various biochemical, physiological, and structural specializations. It is the sum of the processes by which specific metabolic competences are acquired or lost and distinguish daughter cells from each other or from the progenitor cell.

"Morphogenesis" is the integration and coordination of growth and differentiative events occurring at the cellular level and is the process which accounts for the origin of morphological characters and gross form.

Other authors have used the terminology somewhat differently. For example, E. W. Sinnott (1960) in a book entitled *Plant Morphogenesis* wrote, "The process of organic *development, in which are posed the chief problems for the science of morphogenesis, occurs in the great majority of cases as an accompaniment of the process of growth. The association between these two activities (growth and development) is not an invariable one, for there are a few organisms in which growth is completed before development and differentiation are finished, but far more commonly the form and structure of a living thing change while it grows." One example of an exception is the development of the female gametophyte from an 8-nucleate stage in embryo sac development in angiosperms.

Some have employed the term "morphogenesis" in a strictly descriptive sense, essentially as synonymous with classical developmental morphology. More generally and properly, however, it includes, besides descriptive facts as to the origin of form, a study of the results of experimentally controlled development and an analysis of the effects of factors, external and internal, that determine how the development of form proceeds. In other words, it attempts to get at the underlying formativeness in the development of organisms and especially to reach an understanding of the basic fact of which form is the most obvious manifestation, namely, biological organization itself. According to morphogeneticists like Sinnott, "The organism may thus be said to make the cells rather than the cells to make the organism."

F. B. Salisbury and Cleon Ross (1978) used the term "development" (or "morphogenesis") as an inclusive term and regard the phenomenon as consisting of two primary functions: growth and differentiation. They consider growth primarily as an increase in size, and differentiation as the process by which cells become specialized. P. F. Wareing and I. D. J. Phillips (1970) likewise adopted the view that development should be applied in its broadest sense to the whole series of changes which an organism goes through during its life cycle, while noting that it may also be applied to individual organs, to tissues, or even to cells. According to these authors, "plant development" involves both "growth" and "differentiation." "Growth" is used to denote quantitative changes occurring during development and is defined as an irreversible change in the size of a cell, organ, or whole plant. "Differentiation" is applied to qualitative changes. Thus, in their view growth and differentiation are the two major developmental processes. They reserve the term "morphogenesis" as one used by experimental morphologists (morphogeneticists) to denote origin of form. F. C. Steward (1968) employed a quite inclusive connotation of growth. Essentially he used the term "growth" in a very general way to include what others consider more explicitly to be growth, differentiation, and morphogenesis.

With the variable usage of terminology, as has been illustrated, perhaps it is understandable why the coupled terms "growth and development" are used so prevalently. This couplet connotes the kind of concept that James Bonner and A. W. Galston (1952) had when they wrote, "The changing shape, form, degree of differentiation, and state of complexity of the organism constitute the process of development." They viewed "growth" as a quantitative matter concerned with the increasing amount of the organism. On the other hand, "development," in their view, refers to changes in the nature of the growth made by the organism.

Many biologists have emphasized the biological importance of organization and emphasize that the characteristics of life itself are characteristics of a system arising from, and associated with, the organization of materials and processes. It is of utmost importance to keep in mind the fact that there are unique emergent qualities associated with each successively higher level of biological organization—from molecular and subcellular to the levels of cells, tissues, organs, whole organisms, and beyond. In no instance is the cliché that the whole is more than the sum of its parts more vividly exemplified than when we observe the complicated changes which a seed plant manifests during the repeating cycle of development. We can arbitrarily conceive of this cycle as starting with the germination of a seed and continuing with the passage of a juvenile phase of growth and the graduation into maturity. With maturity the organism is capable of shifting from vegetative to reproductive development, with the development of flowers, the development of fruits, and the production again of a new generation of seed. Ultimately, the development of the individual plant ends with senescence and death. This book is concerned with the processes involved in and the mechanisms which control the growth and development of seed plants.

Patterns and Kinetics of Growth in Cells, Tissues, Organs, and Whole Plants

The curve which typically describes the changing size of a growing organism, organ, tissue, cell population, or individual cell is sigmoid in shape (Fig. 1-1a). The sigmoid growth curve can, for convenience, be considered in three parts. First, there is an accelerating phase in which growth starts slowly and gathers momentum. During this period of constantly ac-

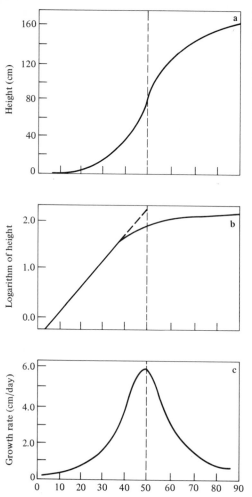

Figure 1-1. Generalized growth curves. **a**, Size-versus-time sigmoid growth curve; **b**, logarithmic growth curve; and **c**, growth rate curve. (Based on data for corn, *Zea mays*, in Whaley, 1961.)

celerating growth rate, increase in size is exponential. When a plant is growing exponentially, it is said to be in the logarithmic phase of its growth. Second, there is either a point of inflection or a phase, more or less protracted, in which the course of growth is approximately linear with time. That is, equal increments of growth tend to occur in equal intervals of time. This is often termed the linear phase of growth. In some cases, however, there is no linear phase but merely a point of inflection in the curve, when the first rising rate of growth gives way to a decreasing rate of growth. Third, there is a phase of declining growth rate until, in fact, growth subsides and the organism may maintain only the size it has already achieved. In annual plants this is followed by senescence and death. In the case of woody perennials, each period of shoot growth is approximately sigmoidal and is followed by dormancy. Shoots of biennials, of course, exhibit two seasons of sigmoidal growth. Instead of plotting merely size versus time (Fig. 1-1a), it is often useful to plot the $\log_e$ of size versus time (Fig. 1-1b) or growth rate versus time (Fig. 1-1c) when analyzing growth. Examples of various specific growth curves and a brief discussion of each follow.

Even the growth of single cells has been demonstrated to be sigmoidal (Fig. 1-2). Organs such as leaves follow a sigmoidal pattern also on the basis of several parameters, including fresh weight, area of lamina, length of lamina, and cell number (Fig. 1-3).

Many fruits exhibit common sigmoid growth curves—e.g., apple (Fig. 1-4), pineapple, strawberry, pea, tomato, etc. However, fruits of many species exhibit more complicated growth curves, which are essentially double sigmoid growth curves. This type of curve is common to probably all the stone fruits such as cherry (Fig. 1-5), apricot, plum, and peach. Such curves are also typical of some nonstone fruits such as fig, grape, and currant. However, neither type of growth curve seems to be distinctive for a particular morphological type of fruit, because there are berries, pomes, and simple and accessory fruits which manifest each.

The double sigmoid growth curve can be explained on the basis of asynchrony of growth of the different parts of the fruit. Two general types of growth centers occur in fleshy fruits, the pericarp and ovule(s). The

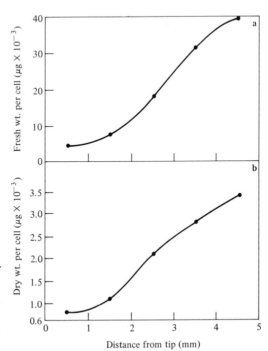

Figure 1-2. Growth curves of single cells in the corn (*Zea mays*) root tip. **a**, Fresh weight; **b**, dry weight. (Redrawn, with permission, from Boss, 1964.)

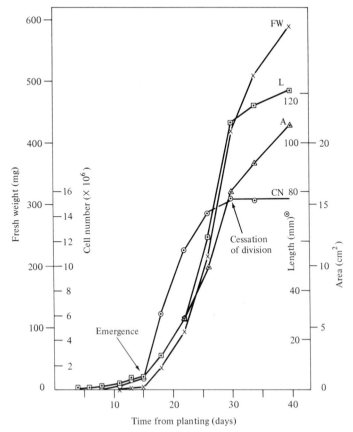

Figure 1-3. Growth curves for leaves of sunflower (*Helianthus annuus*). FW, fresh weight; A, leaf lamina area; L, length; CN, number of cells. (Redrawn, with permission, from Sunderland, 1960.)

growth of the pericarp commonly accounts for the initial enlargement. Growth in later stages generally is associated with seed development.

There is wide variation in the extent to which cell division participates in the growth of fruits, ranging from cases in which cell division (except in the developing seeds) has been completed at the time of pollination (*Ribes, Rubus*), to cases in which there is a brief period of cell division just following pollination (tomato, *Citrus,* cucurbits, apple, *Prunus*), or rather extended periods of cell division (strawberry).

The curve describing the kinetics of growth of whole plants also is typically sigmoidal. Herbaceous annual plants, such as garden pea (*Pisum sativum* L.), exhibit a single sigmoid growth curve throughout ontogeny from germination to senescence and death (Fig. 1-6). Size versus time plots of the shoot growth of woody perennials are sigmoidal throughout ontogeny (Fig. 1-7), although there is a gradual progression toward the

Figure 1-4. Growth curves for apple fruits. **a**, Cell number versus fruit weight; **b**, volume and weight versus time after pollination. (Redrawn from data of Bain and Robertson, 1951, in: *Plant Growth and Development*, 2nd ed., by A. C. Leopold and P. E. Kriedemann. Copyright © 1975 by McGraw-Hill Book Company. Used with permission of McGraw-Hill Book Company.)

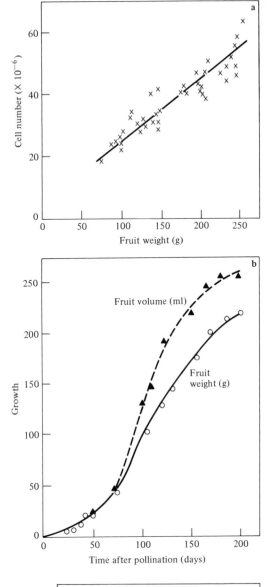

Figure 1-5. Generalized double sigmoid growth curve of cherry. (Redrawn from: *Plant Growth and Development*, 2nd ed., by A. C. Leopold and P. E. Kriedemann. Copyright © 1975 by McGraw-Hill Book Company. Used with permission of McGraw-Hill Book Company.)

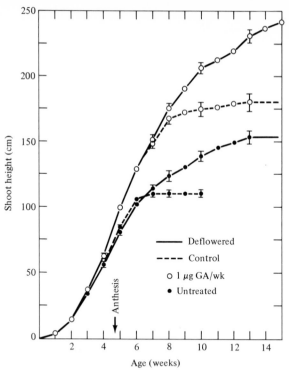

Figure 1-6. Growth curve of a whole herbaceous dicot, pea (*Pisum sativum*). Basic sigmoidicity of the curve is unaffected by removal of flowers or treatment with GA or both. (Redrawn, with permission, from Ecklund and Moore, 1968.)

Figure 1-7. Cumulative secondary stem growth (diameter increase) in several conifers versus age. (Redrawn, with permission, from data of Baker, 1950, in Kramer and Kozlowski, 1960.)

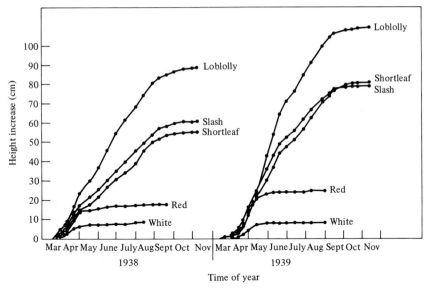

Figure 1-8. Seasonal growth curves (primary shoot or height growth) for several pines for each of two years. (Redrawn, with permission, from data of Kramer, 1943, in Kramer and Kozlowski, 1960.)

asymptote. Notably, however, the size versus time plot for shoot growth of woody plants is composed of increments of growth, each sigmoidal, which alternate with the periods of dormancy. This is true of both primary shoot growth (Fig. 1-8) and secondary shoot growth (Fig. 1-9).

Growth and rate curves for different parameters in woody plants reveal other noteworthy features. For one thing, growth is not synchronous for

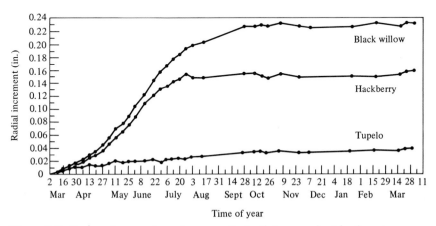

Figure 1-9. Seasonal secondary stem growth (radial expansion) in three tree species. (Redrawn, with permission, from data of Eggler, 1955, in Kramer and Kozlowski, 1960.)

all parts of the plant (Fig. 1-10). Primary shoot growth, or shoot extension, is initiated first by bud break. Production of new leaves, a primary growth process, follows bud break closely. Secondary shoot growth, or increase in diameter, generally follows initiation of primary shoot growth and, in fact, is activated by a hormonal stimulus produced by the growing buds. Root growth of trees tends to be sporadic and to occur whenever physical and chemical environmental factors in the soil and the nutritional status of the shoot permit (Fig. 1-10), true root dormancy apparently being of rare occurrence.

Diurnal variations or periodicity in growth parameters are of common occurrence. Table 1-1 contains data revealing a marked diurnal periodicity in primary shoot growth in two species of *Pinus*.

Seasonal periodicity of shoot growth in trees is well known in temperate regions, indeed at all latitudes where there are distinct seasons. Shoot growth commonly is restricted to a single growing season, and growth alternates with periods of dormancy or rest. In these plants, both the resumption and cessation of shoot growth generally can be correlated with particular physical environmental factors.

However, even in tropical environments trees generally exhibit striking and most intriguing periodicity of shoot growth. An individual shoot exhibits alternate phases of growth and dormancy. During each period of

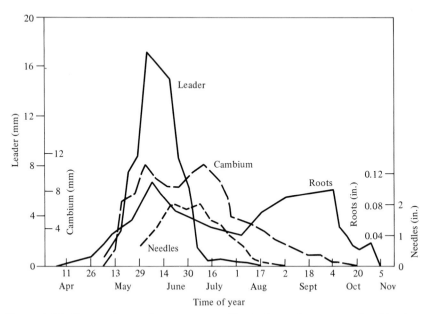

Figure 1-10. Seasonal growth rate curves, illustrating primary growth of leader, secondary growth of stem, primary growth of roots, and growth of new needles. Data are for 10-year-old white pine in southern New Hampshire. (Redrawn, with permission, from data of Kienholz, 1934, in Kramer and Kozlowski, 1960.)

Table 1-1. Diurnal Periodicity of Primary Shoot Growth in Two Species of *Pinus*[a]

	Average day growth (inches)	Average night growth (inches)
Loblolly pine	1.64	3.21
Shortleaf pine	1.27	2.13

[a] Data of Reed (1939) in Kramer and Kozlowski (1960) by permission.

growth new leaves are formed and the young internodes elongate. The terminal bud then enters a state of dormancy until the next growth period begins. Both growing and dormant shoots occur on a single plant, and several "flushes" of shoot growth commonly are exhibited in the course of 1 year.

In tropical regions, where photoperiod is near constant and temperature and other physical environmental factors vary comparatively little, it

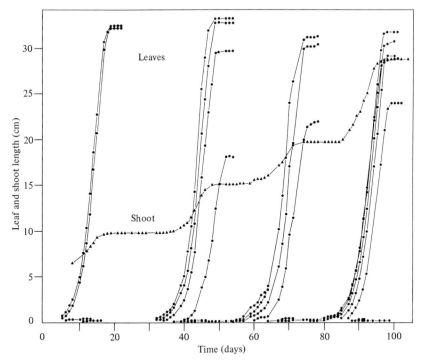

Figure 1-11. Endogenous growth rhythm in the shoot of a tropical tree, *Theobroma cacao,* grown under controlled environmental conditions, expressed as elongation of the shoot and its leaves. The period of shoot-growth rhythm was 26.6 ± 1.1 days measured from the beginning of one dormant period to the beginning of the next. (With permission, from Greathouse et al., 1971.)

appears that growth periodicity is controlled by internal factors that are independent of the physical environment. In other words, the controlling factors of the periodicity of shoot growth in the tropical species are endogenous.

A particularly interesting investigation of endogenous shoot-growth rhythm in tropical species was conducted by Greathouse et al. (1971) with shoot cuttings of *Theobroma cacao*. Shoot cuttings were placed in a growth chamber under constant conditions of a 12-hour photoperiod at a light intensity of 600 fc, a temperature of 29 ± 1°C, and a relative humidity of 70 to 100%. Growth measurements were initiated after the plants had been in the chamber for 4 months, and were repeated at 1- and 2-day intervals for about 100 days. The results were most interesting (Fig. 1-11). There was a remarkable rhythmicity to shoot elongation and leaf production, with a period of 26.6 ± 1.1 days measured from the beginning of one dormant period to the beginning of the next. Furthermore, the plants were not synchronous with regard to shoot growth. They were completely out of phase, as are field-grown *Theobroma cacao* trees in Trinidad. That is to say, the branches were out of phase with each other so far as growth and dormancy were concerned. Dissection of shoot tips revealed further that the total number of leaves and leaf primordia in the shoot apex remained constant during the dormant period and did not increase until the onset of growth. This indicated that activity of the apical meristem, as well as leaf growth and shoot elongation, was rhythmic.

Mathematical Analyses of the Time Course of Growth

Whether growth is regarded as increase in size or however it is measured, it is essentially a process that is quantitative and occurs through time. To the extent that growth is orderly it can be described in mathematical terms. For the purposes of mathematical analysis, growth is usually considered as though it proceeds smoothly and continuously in time, a concept which is not really valid. Our aim now is to show with what success the time course of growth has been mathematically analyzed. Some equations that have been fitted to growth curves will be illustrated (Fig. 1-12), and the validity and applicability of each will be discussed. Shown above the curves are the integrated and linear forms of the equations which describe the changes of size (W) and of the log function of size, respectively, with time.

One equation, which approximately describes growth during the initial phase in many cases is the exponential equation (Fig. 1-12a). This curve depicts a constantly accelerating growth rate and denotes that the rate of change in size is proportional to the size already attained.

Function	Equation	Linear form	W $t = 0$	W $t \rightarrow \infty$
a. Exponential	$W = be^{kt}$	$\log W = \log b + kt$	b	—
b. Monomolecular	$W = a(1 - be^{-kt})$	$\log \dfrac{a - W}{a} = \log b - kt$	$a(1 - b)$	a
c. Logistic	$W = \dfrac{a}{1 + be^{-kt}}$	$\log \dfrac{a - W}{W} = \log b - kt$	$\dfrac{a}{1 + b}$	a
d. Gompertz	$W = ae^{-b \exp(-kt)}$	$\log \log \dfrac{a}{W} = \log b - kt$	ae^{-b}	a
e. Parabola	$W = bt^k$	$\log W = \log b + k \log t$	0	—

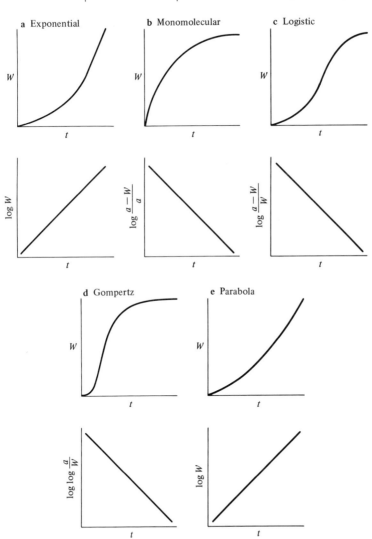

Figure 1-12. Some early equations which are approximate fits for all or part of sigmoid growth curves. (Equations and tabular information from Medawar, 1945, by permission.)

Another is the monomolecular equation (Fig. 1-12b), which describes a curve depicting constantly decelerating growth rate. In essence this equation says that the rate of growth is determined by the amount of growth which is yet to occur. It should be apparent that the monomolecular equation fits approximately the terminal phase of many growth curves.

The logistic equation (Fig. 1-12c) fits approximately many sigmoid growth curves, that is, those curves which consist of an accelerating phase to a point of inflection and a decelerating phase from the inflection point to the asymptotic value which size finally approaches. The equation says that the rate of growth is initially proportional to size, but that the rate decreases as maximum size, which is genetically and environmentally determined, is approached. While approximating a fit of many growth curves, obviously the logistic equation does not fit those curves in which there is a linear phase of growth instead of a mere point of inflection.

A fourth equation which fits growth of woody perennials reasonably well is the Gompertz equation (Fig. 1-12d). This is a "double integral" equation which has maximum flexibility in the fitting of asymmetric sigmoid growth curves. Since the equation describes a gradual progression from the point of inflection to the asymptote, it is sometimes applied to growth curves for trees. The parabolic equation (Fig. 1-12e) approximates only very crudely the initial part of a sigmoid growth curve.

Merits of Mathematical Growth Analysis

Mathematical analysis of growth and growth curves is a convenient, often instructive description of growth, although it tells virtually nothing about the causation or control of growth. If we could devise equations that would accurately predict growth of particular plants or plant parts in given situations, we would have a tool of powerful practical importance. By such means one could, for example, at the seedling stage of a crop, predict the harvest date. Or, more generally, we could predict the amount of growth within any interval. Many workers have arrived at good approximations of equations to do this in particular cases (e.g., citrus trees and fruits in southern California, Turrell et al., 1969). In any case, knowledge of the time course of growth is of utmost importance in numerous cultural practices and in physiological experimentation. It is useful for timing of applications of herbicides and growth regulators, for example, and for predicting time of maximum susceptibility to particular adverse environmental conditions, including pests and pathogens.

Limitations of Mathematical Growth Analysis

No single mathematical statement will describe all types of growth nor perhaps any of them with complete exactness. Growth is a very complex process involving many variables, and it is not to be expected that it can be described by a single equation. Except for the logistic and Gompertz equations, each of the other equations fits only part of a growth curve. It seems that no equation yet devised that expresses a continuous function can satisfactorily describe many growth curves. What is needed are equations expressing discontinuous functions; that is, an equation describing each part of the growth curve. Also, as alluded to earlier, the logistic and Gompertz equations predict a bell-shaped growth rate curve. In many and perhaps most cases of plant growth, however, a linear phase of growth is maintained over a long interval of time. In such a case, there is not a mere point of inflection but a more or less protracted period of constant growth rate. Furthermore, there are some instances in which the typical sigmoid curve does not obtain (e.g., cherry fruit, Fig. 1-5).

Discontinuities in Growth, Growth Periodicities, and Problems of Relative Growth Rate

Mathematical analysis of growth curves assumes that growth occurs continuously and smoothly with time. However, of course there is widespread recognition that growth does not normally proceed at a steady rate, even under uniform environmental conditions, but that it involves certain discontinuities, which sometimes vary sporadically with environmental conditions but frequently are of a cyclical, rhythmic, or periodic nature. Some examples of discontinuities in growth and development will now be cited. The discontinuity in growth by day and by night, as shown by the effects of day length (or night length) and by the contrasted effects of night and day temperature, are evidence that any diurnal cycle is the summation of different effects of day and of night. There is no essential continuity between growth in length and in girth of a plant, and as the growth of a new lateral organ (a bud or a lateral branch, a lateral root, new tillers, adventitious roots) is initiated, new centers are established which become, in large measure, independent. The formation of floral organs, the occurrence of meiosis, and the onset of a sexual or gametophytic phase of growth, syngamy, and later seed development are all stages in development which effectively interrupt the smooth time course of growth and effectively destroy any illusion that it can be uniform in time. The occurrence of storage organs and the phenomena of quiescence and dormancy also show that growth does not bear a uniform, or consistent relation to time and, therefore, any analysis of growth which assumes this must be to this extent inadequate.

There are also problems of relative growth which should be considered when we analyze the growth of whole plants. The concept of relative growth recognizes the unequal growth rates of different parts in one plant body. Examples of different growth rates among the parts of single organs and among different organs of a single plant abound: (1) individual leaflets of a palmately lobed leaf such as that of *Aesculus* (horsechestnut) obviously grow in area and in length at different rates; (2) the shape of dicot leaf lamina such as those of tobacco is determined by the differential growth of the leaf in different parts of the lamina (blade); (3) in particular types of plants, the exponential growth of leaves is arrested with the advancing season, and by a photoperiodic or other stimulus, an underground tuber or storage root develops.

Examples of "growth periodicities" also are abundant. That is to say, developmental cycles are characterized not only by discontinuities but by numerous discontinuities which are cyclical, rhythmic, or periodic. Many plants exhibit a diurnal periodicity in shoot growth, with more shoot elongation occurring at night than during the day (e.g., Table 1-1). During unusually cold nights, however, growth at night may be less than during the day. This behavior was investigated for a number of kinds of plants, most notably tomato in which stem elongation occurs only at night, by F. W. Went some 25 years ago. Noting that many species of plants, particularly those native to temperate regions in which diurnal temperature fluctuations are a characteristic feature of the environment, grow best in an appropriately fluctuating diurnal temperature cycle, Went coined the term "thermoperiodicity" or "thermoperiodism." Tropical plants, which grow normally in regions of little diurnal temperature change, frequently do not exhibit thermoperiodicity but grow best at constant temperatures.

As we have already seen, in perennial plants quiescence and dormancy are conspicuous in temperate latitudes and these phenomena exemplify "seasonal periodicity" in growth and development. Even in tropical climates cycles exist, often in the absence of any apparent correlation with physical environmental stimuli (e.g., Fig. 1-11). The whole range of photoperiodic responses in plant development exemplify periodicities — onset of flowering, onset of dormancy, formation of such organs of perennation as tubers, bulbs, corms, etc. R. C. Friesner showed, as early as 1920, that frequency of mitoses in the roots of pea and corn seedlings showed a diurnal cycle, with more mitoses occurring at night than in the daytime. This rhythm in mitosis and cell division in a root tip is already induced in the seedlings at germination, and, once established, it becomes relatively independent of light and temperature, and tends to continue.

When time is an obvious parameter it is as though organisms have built-in biological clocks. The German physiologist E. Bünning was the first to extensively characterize what we term endogenous circadian rhythms in plants, and they are known for virtually all kinds of organisms (plant and animal) which have been investigated. Circadian rhythm means that the phenomenon recurs at intervals that comprise an approximate 24-hour cycle (e.g., Fig. 1-13), that is, it displays maxima and minima or presence and absence on a cycle of about a 24-hour period. While the nature of the timing mechanism is not known, endogenous circadian rhythms are fascinating and profoundly influence numerous plant re-

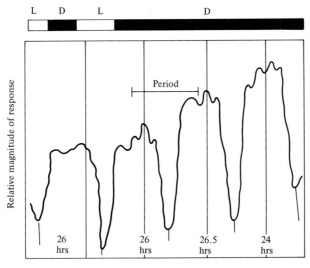

Figure 1-13. Endogenous circadian rhythm of leaf movement in cocklebur (*Xanthium strumarium*). High points indicate high leaf positions (day), troughs low leaf positions (night). Rhythmic leaf movement continues for awhile even in continuous darkness. (Adapted from *Plant Physiology*, 2nd ed., by Frank B. Salisbury and Cleon W. Ross, © 1978 by Wadsworth Publishing Company, Inc., Belmont, California. By permission of the publisher.)

sponses to environmental stimuli and numerous developmental events, e.g., flowering. A remarkable feature about endogenous circadian rhythms is that the length of the period is only slightly influenced by wide variations in temperature, as between 5 and 30°C, even though the rates of metabolic processes upon which these rhythms presumably depend are very much slower at lower temperatures. This type of rhythm obviously has great biological value. Endogenous circadian rhythms enable plants to judge the lengths of the day and night and hence to time reproduction, dormancy, leaf fall, and other activities in relation to the march of the seasons despite erratic changes in temperature in the environment. More is said about endogenous circadian rhythms in Chapter 7.

Mechanisms Controlling Cellular Differentiation

Genetic Basis of Development

That every organism is the product of the interaction of its genetic material and its environment is one of the most fundamental of all biological principles. The genetic material (DNA) determines the potentialities for development. The environment determines the degree and kind of the genetic potential which is actually manifested. Or, to put it a slightly different way, the form and function of plants are dependent upon their genetic makeup and the ability of the genes to express themselves through regulation of protein synthesis. The environment influences the growth, differentiation, and metabolism of a plant by controlling, through effects on both protein synthesis and enzyme activity, expression of its genetic makeup.

Nature of Gene Action

Since the impressive, Nobel prize-winning research of G. W. Beadle and E. L. Tatum in the early 1950s on the mechanism of gene action in *Neurospora* which culminated in the one gene–one enzyme theory, we have understood that the fundamental action of a gene is to control the synthesis of an enzyme or structural protein. Or, in the language of more modern biochemical genetics, the structure of each polypeptide chain is governed by an independent gene or cistron. In the case of several enzymes, known to be made up of two or more polypeptide chains, it is apparent that the structure of each polypeptide chain is governed by a separate independent gene or cistron. Furthermore, we now know that there are functionally different types of genes, and that much of the DNA of a eukaryotic organism acts to control the ultimate expression of the structural genes, which contain the genetic code for specific proteins.

Evidence for Totipotency

There is abundant evidence that at least some and probably all of the kinds of nucleated specialized cells of the mature multicellular organism are in principle "totipotent." A cell which is totipotent contains all the genetic information which was present in the fertilized egg from which it was derived and possesses the capacity to regenerate the adult organism, that is, to behave like a zygote. Hence the term "totipotency" means "total genetic potential."

The intriguing enigma is that all the living cells of the higher plant body potentially have all the attributes of the organism even though, during development, they do not individually display them all. Hence our problem is to try and identify briefly some of the means by which variations from cell to cell of the same individual may have a genetic basis. Current working theory holds that the various types of differentiated cells differ from one another most basically and importantly in the kinds of enzymes and other proteins which they contain and that differentiation therefore is brought about by mechanisms which control the production of the kinds of proteins which each cell contains during its life.

Let us examine initially some of the evidence for totipotency of angiosperm cells. Professor F. C. Steward and associates of the Laboratory for

Cell Physiology, Growth and Development at Cornell University at Ithaca, New York, did some particularly impressive research illustrating the totipotency of higher plant cells. In a series of papers dating from about 1958, Steward et al. first showed that cells derived from the secondary phloem region of mature tap roots of carrot (*Daucus carota*) grow and divide freely when suspended in a liquid medium containing coconut milk (liquid endosperm from coconut seeds or fruits) or liquid endosperm from other sources, and that they do this in various ways that are different from those that obtain in the intact plant body. The freely suspended cells, nourished by coconut milk and a basal medium, dedifferentiate (revert to a meristematic condition) and form loose, randomly growing cultures. It is important to note that if tissue is merely excised and started en masse in a stationary culture on semisolid medium, only callus would form. The loose, randomly growing cultures quickly organize, however, producing first roots, then shoots, to form complete plantlets (Fig. 1-14).

In giving rise to plantlets, a multicellular nodule or growing center is often organized within the proliferated cell culture, which, in its behavior, replaces the zygote or pro-embryo, since it gives rise to a minute embryolike structure which develops entirely from vegetative cells without the stimulus of fertilization. Stages in this growth from free cells strikingly resemble the normal growth of the carrot embryo (Fig. 1-15).

The most important requirement for the cells to grow like embryos instead of merely proliferating to form callus is to start that development from free cells. These early investigations stressed the inherent totipo-

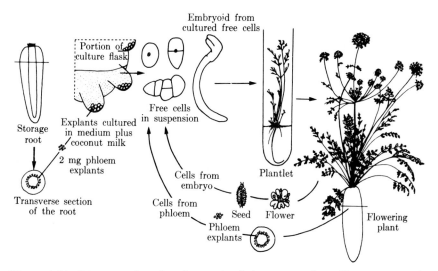

Figure 1-14. Diagram of cycle of growth of the carrot plant (*Daucus carota*), beginning with explants of the secondary phloem region of storage root. (From F. C. Steward, *Growth and Organization in Plants,* © 1968, Addison-Wesley, Reading, Massachusetts. Fig. 10-7. Reprinted with permission.)

crospores) were fully individualized, uninucleate and devoid of starch. Mature pollen grains simply produce pollen tubes but no plantlets. Auxins (IAA or 2,4-D) generally enhanced formation of plantlets but were not essential.

The practical implications of the types of work done by F. C. Steward, J. P. Nitsch, C. Nitsch, and others are very considerable indeed. By means of such techniques, it is theoretically possible to produce large clones of plants with identical genotypes, for example. Breeding programs can be expedited also by methods such as those employed by the Nitsches in that observation of phenotypic expression of a genotype of interest could be observed independent of crossing and the awaiting of seed set, germination, and growth of a particular generation.

Evidence for Differences in Protein Content and for Changes in Protein Content Correlated with Cellular Differentiation

Let us begin with the theory, which has been previously introduced, that cells differ from one another most basically and importantly in the types of proteins (including, of course, enzymes) which they contain and that differentiation is therefore brought about by mechanisms which control the production of proteins. We shall examine first some direct evidence for the first part of the theory, that is, that different kinds of differentiated cells differ qualitatively and quantitatively in protein content. Steward et al. (1965) performed one of the first investigations of this nature. What they sought to do was to determine whether there are detectable differences in the protein complements of different organs of seedling plants, and to ascertain also whether detectable changes occur in the protein complements of the same organs as they develop. They extracted the soluble proteins from 3-day-old pea seedlings (Fig. 1-16) with Tris-glycine buffer ($0.1\,M$, pH 8.3), separated proteins in each extract by electrophoresis on acrylamide gels, stained the proteins, and compared the banding patterns on gels. The studies by Steward et al. revealed quite distinctive differences, as well as obvious similarities, in the protein patterns of different organs of the pea seedling and in root segments at different stages of development (Fig. 1-17). Their study corroborated the results of other investigators, some of whom have monitored changes in activities of various enzymes in the different parts of single organs, e.g., oat coleoptiles and pea roots, during their development.

Primary attention was devoted by Steward et al. to the analysis of soluble proteins in roots. As they pointed out, the growing root is a convenient experimental object for the study of the development of cells from their meristematic to mature state. Successive transverse segments, in a basipetal sequence from the root tip, comprise a series of cell populations

Figure 1-16. Diagram of 3-day-old pea seedling showing parts from which protein extracts were prepared.
1–6, Successive millimeter sections from the root tips;
7, hypocotyl; 8, epicotyl; 9, plumule; and 10, cotyledons. Compare with Fig. 1-17. (Redrawn, with permission, from Steward et al., 1965.)

which represent, in space, a developmental sequence in time. Cells pass
from the meristematic, undifferentiated state in the root tip to specialized
cell types in the more basal parts of the root. As the cells mature in this
way, their protein patterns change, both qualitatively and quantitatively.

Thus, we may conclude that different types of specialized cells in a
multicellular organism exhibit different protein complements in spite of
the fact that all must be presumed, on the basis of other good evidence, to
have the same genetic constitution and therefore the same DNA-encoded
information in their chromosomes. Lest the reader be misled, however, it
should be noted that, while we are now emphasizing the dissimilarities in
protein complements among different kinds of cells, we should appreciate
the probability that the qualitative similarities in protein content greatly
exceed qualitative dissimilarities.

Viewing the evidence just presented, one could postulate, as many biologists have, that the observed differences in protein (enzyme) content
are the cause of differentiation rather than its result. The problem we turn

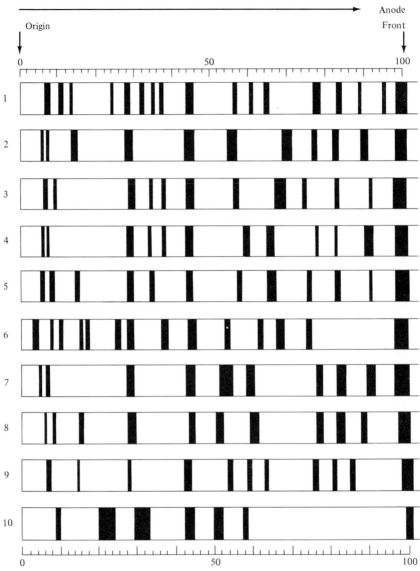

Figure 1-17. Diagrammatic representations of electrophoretic banding patterns seen after electrophoresis of soluble proteins from different parts of 3-day-old pea seedlings. 1–6, Successive millimeter sections from the root tip; 7, hypocotyl; 8, epicotyl; 9, plumule; and 10, cotyledons. Compared with Fig. 1-16. (Redrawn, with permission, from Steward et al., 1965.)

to now is: If all cells in a multicellular organism contain the genetic information necessary for the manufacture of all the different types of proteins found in individuals of that species, what is it that controls the expression of this information? Some subsidiary questions are: (1) How does gene expression result in one protein complement in one part of the plant and an entirely different complement in another part? (2) What is it that controls gene expression such that the same organ can exhibit different protein complements at different developmental stages? (3) Is there direct experimental evidence for selective switching on and off of genes during the course of cellular differentiation and morphogenesis? and (5) Is there any direct evidence for a specific differentiative event being correlated (in a cause and effect manner) with the synthesis of one or more unique proteins?

Mechanisms Controlling Gene Expression

Without going into detail, the salient features of protein synthesis or gene action are:

$$\text{DNA} \xrightarrow{\text{Transcription}} \text{RNA} \xrightarrow{\text{Translation}} \text{Protein} \rightarrow \begin{array}{l}\text{Structure or}\\ \text{Catalysis}\end{array}$$

There is abundant evidence that all genes are not active at all times, that each kind of specialized cell contains its characteristic enzymes, and that each kind of specialized cell produces only a portion of all the enzymes for which its DNA contains information. Major problems now incompletely resolved, however, are: (1) the levels of control of gene expression (of protein synthesis), whether at the transcriptional level or translational level, or both, or others; and (2) the molecular mechanism(s) of control of gene action.

We shall now identify sites of control of gene expression in higher plant systems and describe briefly specific investigations which have provided examples of each.

Sites of Control of Gene Expression

The role of the collective genes of a eukaryotic organism (one whose cells contain definite nuclei, as opposed to prokaryotes such as the bacteria and blue–green algae) is twofold: to control other genes and to control the synthesis of specific structural proteins and enzymes. As previously noted, in the case of enzymes and other proteins which are composed of two or more polypeptide units, the structure of each polypeptide is governed by a separate structural gene.

Consideration of the complex series of events between transcription of a segment of DNA in the formation of a molecule of messenger RNA (mRNA) and the final production of a polypeptide destined to serve as all or part of a structural protein or active enzyme readily reveals that there are many potential sites of control of gene expression. Not surprisingly, it is becoming increasingly apparent that gene expression in eukaryotes does indeed occur at multiple sites, or, to put it another way, control of gene expression occurs by multiple mechanisms in eukaryotes.

Some mechanisms of control of gene expression which have been demonstrated in higher plants are:

1. selective de novo mRNA synthesis or selective gene transcription (transcriptional control);
2. processing and translocation of mRNA (posttranscriptional control);
3. synthesis of new proteins using preformed mRNA (translational control); and
4. activation of preexisting enzymes (posttranslational control).

Another mode of control which has been demonstrated in some animal and primitive plant systems but which has not yet been shown in higher plants is selective gene replication. The clearest, best documented example of this mode of control of gene expression (i.e., synthesis of unique DNA molecules from a specific portion of the total genome) is the amplification of ribosomal RNA genes (rDNA) in oöcytes of amphibians. The extra copies of the rDNA genes are contained in nucleoli and serve as templates for the extraordinary demand for ribosomal RNA synthesis in immature oöcytes. A similar example of rRNA amplification has been reported for the slime mold *Physarum*. Whereas selective gene replication is an important mechanism, genes coding for specialized proteins characteristic of differentiated cells do not appear to be amplified, but rather are present in one or a few copies. Accumulation of gene products in the case of proteins typical of differentiated cells occurs, rather, by reiterated expression of genes (i.e., selective gene transcription).

Selective gene transcription is one major mechanism of control of gene expression. Two examples from higher plants have been particularly well documented. One is the GA-stimulated de novo synthesis of α-amylase and particular other hydrolases in the aleurone layers of cereal grains during early germination. The evidence concerning stimulation of de novo synthesis of α-amylase is the strongest support so far achieved for transcriptional control of a specific enzyme during plant development. This important phenomenon is discussed in detail in Chapter 3. Another example is the stimulation of de novo synthesis of cellulase in pea epicotyls by 2,4-D, a synthetic auxin. In the latter case, putative mRNA was isolated from ribosomes of pea epicotyls after treatment of the epicotyls with 2,4-D. Then the putative mRNA was translated in a cell-free enzyme system from wheat germ, and cellulase was assayed by immunoprecipitation. It was shown that 2,4-D treatment was essential for the synthesis of the mRNA for cellulase. Since the mRNA for cellulase was readily translated in wheat germ, specific factors influencing translation, which might vary among species, appeared not to be involved. Yet, since no delay occurred in mRNA production after 2,4-D treatment but a 24-hour lag period intervened between time of treatment and appearance of cellulase, some form of translational control also appeared to be implicated.

The evidence for other levels of control of gene expression in higher plants comes largely from studies conducted with germinating seeds. Indirect evidence for posttranscriptional control is derived from investigations of enzymic activities which increase during germination, and which are sensitive to inhibitors of protein synthesis but insensitive to inhibitors of RNA synthesis. For many years it has been suggested that some of the protein (enzyme) synthesis that occurs during early germination depends upon mRNA already present in the dry seed and activated for translation during hydration of the seed. Conversion of monoribosomes to polyribosomes and initiation of protein synthesis in the apparent absence of transcription were early observations on which that suggestion was based. More recently there have been efforts to isolate and characterize a specific mRNA fraction from ungerminated seeds and to demonstrate its role in protein synthesis.

There is some evidence for posttranscriptional control from investigations conducted by L. Dure and associates with cotton seed. Major reserves stored in the cotyledons of a cotton seed are proteins and fats. Two enzymes which play key roles in the breakdown of these reserves are carboxypeptidase C and isocitrate lyase. The former degrades proteins, and the latter enzyme is needed in the glyoxylate cycle for utilization of the stored fat. Carboxypeptidase C and isocitrate lyase activities are not detectable in the dry cotton seed but activity appears in the cotyledons during the second day of germination. The appearance of both enzymic activities is inhibited by cycloheximide (inhibitor of cytoplasmic protein synthesis) but not by actinomycin D (inhibitor of RNA synthesis). Moreover, it has been shown that carboxypeptidase C is synthesized de novo from constituent amino acids. This and other evidence may indicate that the de novo synthesis of carboxypeptidase C depends upon translation of preformed mRNA which is already present in the dry cotton seed prior to the initiation of seed germination, although this point at present is uncertain.

This brief discussion should serve to illustrate that the appearance of enzyme activities (hence gene expression) is regulated at several different levels, concomitantly in part and sequentially in part, during the development of higher plant systems. For an excellent recent review on the regulation of RNA metabolism by auxins, gibberellins, and cytokinins, the article by John V. Jacobsen (1977) should be consulted.

Introduction to Plant Hormones

Development obviously is a rigidly regulated process. One has but to contemplate the remarkable structural and functional unity of the plant body throughout ontogeny and the chronology of developmental events that occur in individuals of each species to recognize this fact.

The plant hormones are extremely important agents in the integration of developmental activities, and they also are concerned importantly in the response of plants to the external physical environment. Environmental factors often exert inductive effects by evoking changes in hormone metabolism and distribution within the plant. Aside from participation in responses to inductive environmental effects, hormones also are the principal agents which otherwise regulate expression of the intrinsic genetic potential of plants.

The literature on plant hormones is replete with variable attempts to define or describe these important substances. Most so-called definitions are of the operational type; that is, they define hormones in terms of their biological activity, often to the virtual exclusion of any mention of their chemical nature. An example is the definition which the author has adopted for plant hormones or phytohormones generally.

A plant hormone may be defined—perhaps described would be more accurate—as an organic substance other than a nutrient (a substance which supplies either carbon and energy or essential mineral elements), active in very minute amounts (e.g., <1 mM, often <1 μM), which is formed in certain parts of the plant and which usually is translocated to other sites, where it evokes specific biochemical, physiological, and/or morphological responses. Plant hormones are active in tissues where they are produced as well as at a distance. A growth regulator (= growth substance or plant growth regulator) is an organic compound (other than a nutrient) which in small amounts (<1 mM) (or low concentrations) promotes, inhibits, or qualitatively modifies growth and development. Growth inhibitors are organic compounds which retard growth generally and which have no stimulatory range of concentrations. Hence, all hormones (natural plant products) are growth regulators, but the converse is not true. There are literally hundreds of purely synthetic compounds which qualify as growth regulators but which are not hormones.

The commonly recognized classes of plant hormones are the auxins, gibberellins, cytokinins, abscisic acid and other growth inhibitors, ethylene, and the hypothetical florigens or anthesins. In the following chapters we shall devote attention to the following topics, where we can, about each class of hormones: definitions or descriptions; history of discovery; chemical characterization; occurrence; biosynthesis, enzymic degradation, and other features of metabolism; transport; physiological effects; and mechanisms of action. In a final chapter we shall devote extensive attention to phytochrome and its role in numerous photoresponses and the roles of hormones in flowering.

References

Avery, G. S., Jr. 1933. Structure and development of the tobacco leaf. *Am. J. Bot.* **20**: 565–592.

Bergmann, L. 1960. Growth and division of single cells of higher plants *in vitro.* *J. Gen. Physiol.* **43**: 841–851.

Black, M. and J. Shuttleworth. 1976. Inter-organ effects in the photocontrol of growth. In: Smith, H., ed. *Light and Plant Development*. Butterworth & Company, Limited, London. Pp. 317–331.

Bonner, J. 1965. *The Molecular Biology of Development*. Oxford University Press, New York.

Bonner, J. and A. W. Galston. 1952. *Principles of Plant Physiology*. W. H. Freeman and Company, San Francisco.

Boss, M. L. 1964. Equational relations between nutrient quantities and cellular growth. *New Phytol.* **63**: 47–54.

Britten, R. J. and E. H. Davidson. 1969. Gene regulation for higher cells: a theory. *Science* **165**: 349–357.

Bünning, E. 1977. Fifty years of research in the wake of Wilhelm Pfeffer. *Ann. Rev. Plant Physiol.* **28**: 1–22.

Carr, D. J. and K. G. M. Skene. 1961. Diauxic growth curves of seeds with special reference to French beans (*Phaseolus vulgaris* L.). *Aust. J. Biol. Sci.* **14**: 1–12.

Dure, L. S., III. 1975. Seed formation. *Ann. Rev. Plant Physiol.* **26**: 259–278.

Ecklund, P. R. and T. C. Moore. 1968. Quantitative changes in gibberellin and RNA correlated with senescence of the shoot apex in the 'Alaska' pea. *Am. J. Bot.* **55**: 494–503.

Eggler, W. A. 1955. Radial growth in nine species of trees in southern Louisiana. *Ecology* **36**: 130–136.

Erickson, R. O. 1976. Modeling of plant growth. *Ann. Rev. Plant Physiol.* **27**: 407–434.

Evans, G. C. 1972. *The Quantitative Analysis of Plant Growth*. University of California Press, Berkeley.

Friesner, R. C. 1920. Daily rhythms of elongation and cell division in certain roots. *Am. J. Bot.* **7**: 380–407.

Galston, A. W. and P. J. Davies. 1970. *Control Mechanisms in Plant Development*. Prentice-Hall, Englewood Cliffs, New Jersey.

Gibbs, J. L. and D. K. Dougall. 1963. Growth of single plant cells. *Science* **141**: 1059.

Greathouse, D. C., W. M. Laetsch, and B. O. Phinney. 1971. The shoot-growth rhythm of a tropical tree, *Theobroma cacao. Am. J. Bot.* **58**: 281–286.

Gross, P. R. 1968. Biochemistry of differentiation. *Ann. Rev. Biochem.* **37**: 631–660.

Haber, A. H., T. J. Long, and D. E. Foard. 1964. Is final size determined by rate and duration of growth? *Nature (London)* **201**: 479–480.

Heslop-Harrison, J. 1967. Differentiation. *Ann. Rev. Plant Physiol.* **18**: 325–348.

Heslop-Harrison, J. 1972. Genetics and the development of higher plants: a summary of current concepts. In: Steward, F. C., ed. *Plant Physiology. Vol. VIC. Physiology of Development: From Seeds to Sexuality*. Academic Press, New York. Pp. 341–366.

Higgins, T. J. V., J. A. Zwar, and J. V. Jacobsen. 1976. Gibberellic acid enhances the level of translatable mRNA for α-amylase in barley aleurone layers. *Nature (London)* **260**: 166–168.

Jacobsen, J. V. 1977. Regulation of ribonucleic acid metabolism by plant hormones. *Ann. Rev. Plant Physiol.* **28**: 537–564.

Kienholz, R. 1934. Leader, needle, cambial and root growth of certain conifers and their relationships. *Bot. Gaz.* **96**: 73–92.

Kramer, P. J. 1943. Amount and duration of growth of various species of tree seedlings. *Plant Physiol.* **18**: 239–251.

Kramer, P. J. and T. T. Kozlowski. 1960. *Physiology of Trees.* McGraw-Hill Book Company, New York.

Leopold, A. C. and P. E. Kriedemann. 1975. *Plant Growth and Development.* 2nd ed. McGraw-Hill Book Company, New York.

Lockhart, J. A. 1965. An analysis of irreversible plant cell elongation. *J. Theoret. Biol.* **8**: 264–275.

Loomis, W. E. 1934. Daily growth of maize. *Am. J. Bot.* **21**: 1–6.

Medawar, P. B. 1945. Size, shape, and age. In: Clark, W. E. L. and P. B. Medawar, eds. *Essays on Growth and Development.* Oxford University Press, Oxford. Pp. 157–187.

Meins, F., Jr. 1975. Cell division and the determination phase of cytodifferentiation in plants. In: *Results and Problems in Cell Differentiation. Vol. 7. Cell Cycle and Differentiation.* Springer-Verlag, Berlin. Pp. 151–175.

Millerd, A. 1975. Biochemistry of legume seed proteins. *Ann. Rev. Plant Physiol.* **26**: 53–72.

Nelson, O. E. and B. Burr. 1973. Biochemical genetics of higher plants. *Ann. Rev. Plant Physiol.* **24**: 493–518.

Nitsch, J. P. and C. Nitsch. 1969. Haploid plants from pollen grains. *Science* **163**: 85–87.

Plaisted, P. H. 1957. Growth of the potato tuber. *Plant Physiol.* **32**: 445–453.

Richards, F. J. 1969. The quantitative analysis of growth. In: Steward, F. C., ed. *Plant Physiology. Vol. VA. Analysis of Growth: Behavior of Plants and Their Organs.* Academic Press, New York, Pp. 3–76.

Salisbury, F. B. and C. W. Ross. 1978. *Plant Physiology.* 2nd ed. Wadsworth Publishing Company, Belmont, California.

Sangwan, R. S. and B. Norreel. 1975. Induction of plants from pollen grains of *Petunia* cultured *in vitro. Nature (London)* **257**: 222–224.

Sinnott, E. W. 1960. *Plant Morphogenesis.* McGraw-Hill Book Company, New York.

Steeves, T. A. and I. M. Sussex. 1972. *Patterns in Plant Development.* Prentice-Hall, Englewood Cliffs, New Jersey.

Steward, F. C. 1968. *Growth and Organization in Plants.* Addison-Wesley Publishing Company, Reading, Massachusetts.

Steward, F. C. 1970. Totipotency, variation and clonal development of cultured cells. *Endeavour* **29**: 117–124.

Steward, F. C. and A. D. Krikorian. 1971. *Plants, Chemicals and Growth.* Academic Press, New York.

Steward, F. C., R. F. Lyndon, and J. T. Barber. 1965. Acrylamide gel electrophoresis of soluble plant proteins: a study on pea seedlings in relation to development. *Am. J. Bot.* **52**: 155–164.

Steward, F. C., E. M. Shantz, J. K. Pollard, M. O. Mapes, and J. Mitra. 1961. Growth induction in explanted cells and tissues: metabolic and morphogenetic manifestations. In: Rudnick, D., ed. *Synthesis of Molecular and Cellular Structure.* The Ronald Press, New York. Pp. 193–246.

Sunderland, N. 1960. Cell division and expansion in the growth of the leaf. *J. Exp. Bot.* **11**: 68–80.

Sweeney, B. M. 1969. *Rhythmic Phenomena in Plants*. Academic Press, New York.

Thimann, K. V. 1977. *Hormone Action in the Whole Life of Plants*. University of Massachusetts Press, Amherst.

Torrey, J. G. 1967. *Development in Flowering Plants*. Macmillan Company, New York.

Trewavas, A. 1976. Post-translational modification of proteins by phosphorylation. *Ann. Rev. Plant Physiol.* **27**: 349–374.

Tukey, H. B. 1934. Growth of the embryo, seed, and pericarp of the sour cherry (*Prunus cerasus*) in relation to season of fruit ripening. *Proc. Am. Soc. Hort. Sci.* **31**: 125–144.

Turrell, F. M., M. J. Garber, W. W. Jones, W. C. Cooper, and R. H. Young. 1969. Growth equations and curves for citrus trees. *Hilgardia* **39**: 429–445.

Verma, D. P. S., G. A. Maclachlan, H. Byrne, and D. Ewings. 1975. Regulation and *in vitro* translation of messenger ribonucleic acid for cellulase from auxin-treated pea epicotyls. *J. Biol. Chem.* **250**: 1019–1026.

Vince-Prue, D. 1975. *Photoperiodism in Plants*. McGraw-Hill Book Company, New York.

Wareing, P. F. and I. D. J. Phillips. 1970. *The Control of Growth and Differentiation in Plants*. Pergamon Press, New York.

Went, F. W. 1957. *The Experimental Control of Plant Growth*. Ronald Press, New York.

Whaley, W. G. 1961. Growth as a general process. In: Ruhland, W., ed. *Handbuch der Pflanzenphysiologie*. Vol. XIV. Springer-Verlag, Berlin. Pp. 71–112.

Wilkins, M. B. 1969. Circadian rhythms in plants. In: Wilkins, M. B., ed. *Physiology of Plant Growth and Development*. McGraw-Hill Publishing Company Limited, London. Pp. 647–671.

Wilson, C. C. 1948. Diurnal fluctuations of growth in length of tomato stem. *Plant Physiol.* **23**: 156–157.

Ycas, M., M. Sugita, and A. Bensam. 1965. A model of cell size regulation. *J. Theoret. Biol.* **9**: 444–470.

Zimmermann, M. H. and C. L. Brown. 1971. *Trees—Structure and Function*. Springer-Verlag, Berlin.

Chapter 2

Auxins

Brief History of Discovery

The first type of plant hormone to be discovered was the auxins. The term "auxin" is derived from the Greek "auxein" which means "to grow," and was proposed originally by Kögl and Haagen-Smit and by F. A. F. C. Went to designate a particular substance which had the property of promoting curvature in the *Avena* Coleoptile Curvature Test, which will be described later.

Devising an all inclusive and exclusive definition of auxins is difficult, as is the case of all other plant hormones. A committee of plant physiologists in 1954 defined "auxin" as a generic term for compounds characterized by their capacity to induce elongation in shoot cells. It was emphasized that auxins resemble indole-3-acetic acid, the only known naturally occurring auxin, in physiological action and that they affect other processes besides elongation, even though elongation is considered critical. Years later K. V. Thimann (1969) defined "auxins" as "organic substances which at low concentrations ($<0.001\ M$) promote growth (cell enlargement) along the longitudinal axis, when applied to shoots of plants freed as far as practical from their own inherent growth-promoting substance, and inhibit the elongation of roots." Some authors have restricted the term to substances active in the *Avena* Coleoptile Curvature Test.

A very extensive discussion would be required to recount in detail the discovery, isolation, and eventual identification of the well-established natural auxin, indoleacetic acid. We can only describe the highlights here. One could say that the development of the hormone concept in general had its beginning as long ago as 1758 when Duhamel du Monceau, a Frenchman, conducted experiments which led him to conceive of growth correlations in plants brought about by two saps, one moving downward, the other upward. But, actually, the German botanist Julius von Sachs, known as the Father of Plant Physiology, is credited in 1880 with having launched the first theory of substances, which we now call hormones, controlling plant growth. Sachs envisaged the existence of organ-forming substances moving in various polar patterns and controlling growth and development. His thesis was that morphological differences among the different organs are due to corresponding differences in their material composition, which must be already present at the time of initiation, even though in his time chemical reactions and other crude methods failed to show any difference. Sach's theory found little support among his contemporaries, and it was nearly half a century before the existence of a substance corresponding to his organ-forming substances was finally proved. And, in fact, to suggest that any kind of plant hormone is a specific organ-forming substance is misleading in view of our present knowledge that each kind of hormone influences multiple developmental processes.

The discovery of auxins was the outcome of experiments on phototropism and, to a lesser extent, geotropism. Charles Darwin, better known generally for his theory of evolution, was the first investigator in the field of auxinology. Darwin was interested in plant movements, especially phototropism, and his observations, published in 1880 in a delightful book entitled *The Power of Movement in Plants,* led ultimately to the definitive discovery of auxin. Darwin performed very simple experiments on the phototropism of coleoptiles (Fig. 2-1) of canary grass (*Phalaris*) seedlings. When coleoptiles of etiolated canary grass seedlings were illuminated from one side, a strong positive curvature (positive phototropism, growth toward the light source) resulted. If the tip of the coleoptile was darkened by a tin foil cap and only the lower part was unilaterally illuminated, curvature usually did not result. However, when he reversed this procedure, that is, when only the lower part was darkened and the coleoptile tip was exposed, curvature resulted. Darwin also demonstrated that a coleoptile does not react phototropically when 2.5 to 4 mm of the tip is removed. From these empirical experiments, Darwin concluded "that when seedlings are freely exposed to a lateral light, some influence is transmitted from the upper to the lower part, causing the latter to bend."

Several other investigators extended Darwin's experiments, but it remained for a young Dutch botany student in Utrecht named Frits W. Went to make the definitive discovery of auxin in 1926–28. Went was, at the time, serving in the Dutch army, but at night he would go to his father's (F. A. F. C. Went) plant physiology laboratory at the university to keep up his graduate studies in botany. Went's favorite object of study was the coleoptile of oat (*Avena sativa*) (Fig. 2-1). The coleoptile of *Avena sativa* starts its development as a dome-shaped structure which covers the shoot apex. Cell divisions cease early in the life of the coleop-

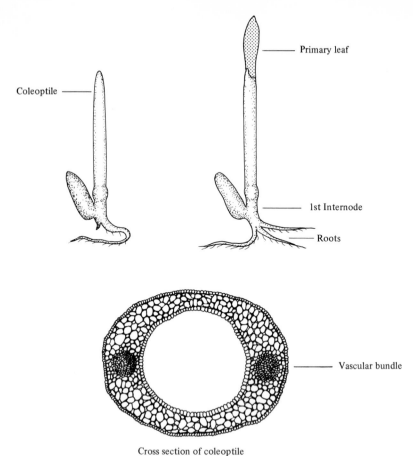

Cross section of coleoptile

Figure 2-1 The *Avena* coleoptile. The coleoptile is a hollow cylinder enclosing the epicotyl and is attached to the axis of the seedling at the first node. It is a leaf-like structure and the first part of the plant to emerge from the soil. Coleoptiles are characteristic of the grasses. The first leaf, as it expands, eventually pierces the coleoptile at its apex after which growth of the coleoptile ceases. Cell division takes place in an etiolated *Avena* coleoptile until it reaches about 1 cm in length, after which cell division ceases and it elongates for 4 days at a rate of approximately 1 mm/hour. (Redrawn, with permission, from Bonner and Galston, 1952.)

tile, and it enlarges principally by cell elongation for the first 70 to 100 hours of seedling growth. During most of its growth period, elongation of the coleoptile takes place primarily in the center of the organ, almost to the exclusion of the extreme apical end and basal portion. Under environmental conditions usually employed, oat coleoptiles reach a height of 3 cm in light and 6 cm in darkness and at maturity are about 1.5 mm in diameter.

Went's basic experimental design really was quite simple. Tips were

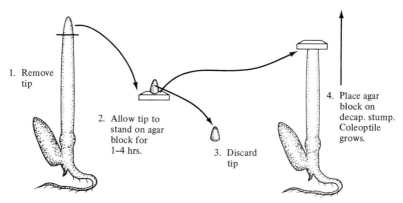

1. Remove tip

2. Allow tip to stand on agar block for 1–4 hrs.

3. Discard tip

4. Place agar block on decap. stump. Coleoptile grows.

Figure 2-2. Basic experiment by F. W. Went which led to the definitive discovery of auxin. (Redrawn, with permission, from Bonner and Galston, 1952.)

excised from etiolated oat coleoptiles and placed uniformly on a thin layer of 3% agar. After 1 to 4 hours, the tips were removed and the agar was sliced into small blocks. The small blocks of agar were then placed upon the stumps of decapitated coleoptiles. The result was that the rate of elongation of the coleoptiles was promoted just as if the stumps had been capped with fresh coleoptile tips (Fig. 2-2). By this simple, if tedious, procedure the definitive discovery of auxin was made and was first published in 1928. Reportedly it was 3:00 AM on the morning of April 17, 1926 that F. W. Went saw the successful result of his experiment. He dashed home and awakened his father to report the news. His father bawled him out for awakening him, and told Frits he would look at his results the next morning!

Went's *Avena* Coleoptile Curvature Test

As noted previously the growth-promoting property of coleoptile tips, or of agar blocks which have been in physical contact with them, may be demonstrated in a striking manner by applying the tip or agar block asymmetrically (one-sidedly) to the cut surface of a decapitated coleoptile stump. Under these conditions the growth-promoting substance passes down the coleoptile in an asymmetric flux, causing more growth to occur on the side of the stump subtending the source of auxin than on the other, resulting in a curvature of the coleoptile. Since curvature is away from the applied block or tip, it is referred to as a negative curvature. The angle of curvature is proportional, over a wide range, 0 to 20°, to the concentration of auxin in the agar block.

F. W. Went proceeded to devise a bioassay for auxin, the *Avena* Coleoptile Curvature Test, often called simply the *Avena* test, which re-

Figure 2-3. Procedures in the *Avena* Coleoptile Curvature Test. (Redrawn, with permission, from Leopold, 1955.)

mains the most sensitive and specific bioassay for auxins. A bioassay is, of course, a biological testing procedure according to which the response of a plant or plant part is used to detect the presence of an active substance and to measure the amount present. The essential steps of the *Avena* Coleoptile Curvature Test are outlined in Fig. 2-3.

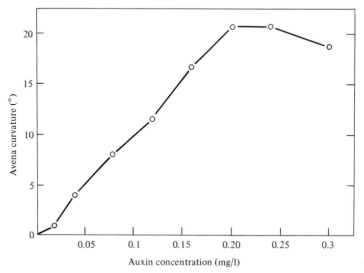

Figure 2-4. Typical dose–response curve for the *Avena* Coleoptile Curvature Test, showing direct proportionality of degrees of negative curvature to arithmetic auxin concentration, up to about 0.2 mg/ liter. (Redrawn, with permission, from Leopold, 1955.)

In the *Avena* Coleoptile Curvature Test, response (negature curvature) is linearly proportional to arithmetic concentration of auxin (up to about 0.20 ppm) (Fig. 2-4). However, in other bioassays (e.g., *Avena* straight growth test), responses are linearly proportional to logarithm of concentration, over specific ranges of concentration.

Early Isolations of IAA

Although Went had succeeded in isolating auxin by a diffusion technique, he was not able to isolate it in pure form or to identify the substance chemically. The first attempt to isolate auxin, in sufficient quantity to permit a chemical analysis of its structure, was not carried out on oat coleoptiles or indeed even on plant tissue, as it occurs there in much too small quantities for the analytical methods available at that time. Biochemists in their searches for sources of substances active in the *Avena* test found other biological materials which were far richer in substances active in the *Avena* test. Kögl and Haagen-Smit (1931) at Utrecht, Holland, obtained an active substance called "auxin A" (auxentriolic acid) from 33 gal of human urine. Kögl, Haagen-Smit, and Erxleben (1934) analyzed urine again and found substances called "auxins A and B" + "heteroauxin." On subsequent analysis heteroauxin was found to be identical with indole-3-acetic acid (IAA) (Fig. 2-5). Kögl and Kostermans (1934) isolated IAA from yeast plasmolysate. K. V. Thimann isolated IAA from large-scale cultures of the fungus *Rhizopus suinus* in 1935. The first generally accepted report of the occurrence of IAA in a higher plant was published by A. J. Haagen-Smit et al. in 1946. Since then there have been reports of the occurrence of IAA in dozens of species of seed plants. It is now generally agreed that indoleacetic acid is the major, perhaps the only, native auxin of higher plants. Interestingly, IAA had long been known to chemists since it was first synthesized in Germany in 1904 by Ellinger, but never before had it been suspected of having biological activity.

Synthetic Auxins

There are many purely synthetic compounds which may correctly be called auxins (Fig. 2-6), since they exhibit physiological action similar to that of IAA. While these compounds are auxins, of course they technically may not be called hormones.

Synthetic auxins are chemically diverse, but may for convenience be

Figure 2-5. Structural formula of indole-3-acetic acid.

Indole acids

Indolepropionic acid

Indolebutyric acid

Naphthalene acids

Naphthaleneacetic acid   β–Naphthoxyacetic acid

Chlorophenoxy acids

(2, 4–D)   (2, 4, 5–T)   (MCPA)

2, 4,–Dichlorophenoxyacetic acid   2–Methyl–4–chlorophenoxyacetic acid

2, 4, 5–Trichlorophenoxyacetic acid

Benzoic acids

2, 4, 6–Trichloro–   2, 3, 6–Trichloro–
benzoic acid   benzoic acid

2–Methoxy–3, 6–dichlorobenzoic acid (dicamba)

Picolinic acids

4–Amino–3, 5, 6–trichloropicolinic acid

[Tordon (picloram)]

Figure 2-6. Examples of synthetic auxins and auxin-type herbicides.

categorized in five major groups: indole acids, naphthalene acids, chlorophenoxy acids, benzoic acids, and picolinic acid derivatives. Examples of
the first group are indolepropionic acid and indolebutyric acid. However,
these two compounds apparently are not exclusively synthetic, since each
has been reported to occur naturally in a few species. The second group is
typified by naphthaleneacetic acid and β-naphthoxyacetic acid, both of

which are synthetic auxins of long standing. Among the synthetic auxins in the chlorophenoxy acid series, the best known are 2,4-dichlorophenoxyacetic acid (2,4-D), 2,4,5-trichlorophenoxyacetic acid (2,4,5-T), and 2-methyl-4-chlorophenoxyacetic acid (MCPA). Several true auxins which promote growth in a manner like IAA at physiological concentrations are best known as selective herbicides when used at sufficiently high concentrations. More will be said about the chlorophenoxy auxins in the succeeding section.

Common synthetic auxins in the benzoic acid group are 2,3,6- and 2,4,6-trichlorobenzoic acids and dicamba, the latter being a powerful herbicide. It is effective against some species of deep-rooted perennials, e.g., wild morning glory (*Convolvulus arvenis*) and Canada thistle (*Cirsium arvense*), which are not readily killed by 2,4-D.

Among the best known of the picolinic acid series is the auxin-type herbicide known as picloram or Tordon (4-amino-3,5,6-trichloropicolinic acid), which is one of the most powerful selective herbicides known.

Controversy Surrounding the Use of Certain Chlorophenoxy Acids as Herbicides and Defoliants

Background

Among the auxin-type herbicides, certain chlorophenoxy acids are among the oldest known, most selective as to species affected, and most effective. Best known among compounds of this type are 2,4-dichlorophenoxyacetic acid (2,4-D), 2,4,5-trichlorophenoxyacetic acid (2,4,5-T), and 4-chloro-2-methylphenoxyacetic acid (MCPA). Research on these compounds was conducted in secrecy during World War II because of their potential use as agents of biological warfare, and they became available for commercial use in the late 1940s.

2,4-D, the first of the organic selective herbicides to be discovered and applied widely, and 2,4,5-T, the two compounds to which further discussion will be restricted, are both very active auxins. At concentrations equivalent to physiological concentrations of IAA, they are active in most auxin bioassays and are widely used instead of the native auxin IAA in experimentation. This is because they are less rapidly destroyed by the IAA oxidase system than IAA and less readily form conjugates, e.g., with amino acids. Incidentally, it should also be noted that 2,4,-D and 2,4,5-T exhibit only weak polar transport, as compared to IAA.

2,4-D and 2,4,5-T are very potent and highly selective herbicides, being lethal in adequate concentration to most dicotyledonous (broad-leaved) plants and comparatively ineffective on most monocotyledonous plants.

However, it is to be stressed that selectivity in herbicidal action is relative, as regards species specificity, as well as of course a function of herbicide concentration. Hence, while 2,4-D will kill many dicotyledonous species at concentrations which are harmless to some monocots, at
sufficiently high concentrations it will also kill monocot species.

2,4,5-T is used in a variety of commercial formulations and in recent
years this herbicide has received major attention. The formulations used
in commercial sprays include the free acid, salts, and a variety of esters
and amine salts. The free acid form of 2,4,5-T is practically insoluble in
water; the sodium salt is only soluble to a limited extent (<3%). The
amine salts are considerably more soluble in water, but they are somewhat difficult to prepare. Ester formulations are most commonly used as
oil–water emulsions. Esters and amine salts are most commonly used.
Because of the problem of drift through air and damage to nontarget species, highly volatile esters are no longer commercially available. 2,4,5-T is
more expensive than 2,4-D, hence it has been used primarily to control
certain woody plants (e.g., sagebrush and mesquite in the western United
States) and a few herbaceous species on which it is more effective than
2,4-D. Because of the cost difference, commercial formulations containing 2,4,5-T usually are mixtures of the two compounds. The *n*-butyl ester
of 2,4,5-T in a 1:1 mixture with 2,4-D is known as ''Agent Orange,'' which
the United States military used extensively as a defoliant in Vietnam during the United States' involvement in the recent conflict between North
and South Vietnam. The spray concentrations usually vary between 0.1
and 2.5% and the rates of application are usually between 0.5 and 8 lb per
acre. Higher rates and concentrations were used in Vietnam for military
purposes.

Chlorinated dibenzo-*para*-dioxins have been recognized as possible
by-products in manufacturing certain chlorinated phenols since 1950. As
many as 75 chlorodioxins are theoretically possible, but only a few have
been isolated and investigated. One very prominent exception is 2,3,7,8-
tetrachlorodibenzo-*para*-dioxin (TCDD) (Fig. 2-7), which is produced as a
by-product in the manufacture of 2,4,5-T. TCDD has been called the most
toxic synthetic chemical known. Perhaps it will be of interest to examine
the procedures for the chemical synthesis of 2,4,5-T and to show where
and how TCDD is formed.

TCDD   2, 4, 5–T

Figure 2-7. Structural formulas of 2,3,7,8-tetrachlorodibenzo-*para*-dioxin (TCDD)
and 2,4,5-trichlorophenoxyacetic acid (2,4,5-T).

Chemical Synthesis of 2,4,5-T[1]

The usual starting material for the chemical synthesis of 2,4,5-T is 1,2,4,5-tetrachlorobenzene which can be reacted with methanol and sodium hydroxide under high temperature and pressure conditions for several hours to produce the sodium salt of 2,4,5-trichlorophenol (Reaction 1):

1, 2, 4, 5–Tetrachloro-   2, 4, 5–Trichloroanisole   Sodium salt of
benzene   2, 4, 5–Trichlorophenol

2,4,5-Trichloroanisole is thought to be an intermediate in this reaction. The high temperature and pressure and strong alkalinity required for Reaction 1 must be critically selected and rigorously controlled during a finite reaction time in order to minimize side reactions and the occurrence of impurities in the final product. The impurities present in any commercial preparation of 2,4,5-T depend strongly on the purity of the starting materials and the reaction conditions.

The aqueous trichlorosodium phenoxide is next reacted with chloroacetic acid under mildly alkaline conditions (Reaction 2):

This product is then acidified with H_2SO_4 to produce 2,4,5-T (Reaction 3):

The conditions for Reactions 2 and 3 are mild compared with those required for the hydrolysis in Reaction 1. TCDD results primarily as a side product of Reaction 1, but also apparently from impurities in the starting materials.

Unfortunately, formulations of 2,4,5-T which were commercially available prior to about 1968 were contaminated with from 2 to 50 parts per

[1] From: Report on 2,4,5-T. A report of the Panel on Herbicides of the President's Science Advisory Committee, Executive Office of The President, Office of Science and Technology, March 1971. 68 pp.

million (ppm) of TCDD. Commercial preparations of 2,4,5-T available in the United States are now legally required to contain 0.1 ppm or less of TCDD.

Toxicities of 2,4,5-T and TCDD

Initial concern for the possible hazard to humans and other animals exposed to 2,4,5-T came from research conducted by the Bionetics Research Institute under contract with the National Cancer Institute of the U. S. Public Health Service, circa 1964 to 1968. In those studies it was found that mice and rats treated during early pregnancy with large doses of 2,4,5-T gave birth to defective offspring. Announcement of those results, together with reports of alleged increased occurrence of human birth defects by South Vietnamese newspapers in June and July 1969, elicited prompt and far–reaching reactions from governmental agencies, segments of the scientific community, various citizens' groups, and public mass communications media.

The concern was so great that in the United States, by joint order of the Secretaries of Agriculture and of Health, Education and Welfare, the use of 2,4,5-T in the United States was curtailed for a time to the point that, practically speaking, it could not be used. Two major manufacturers of 2,4,5-T in the United States, Dow Chemical Company and Hercules Incorporated, exercised their legal rights to petition for referral of the matter to an Advisory Committee. Such an Advisory Committee subsequently was appointed from members of the National Academy of Sciences, and a U. S. government bulletin entitled *Report on 2,4,5-T— A Report of the Panel on Herbicides of the President's Science Advisory Committee* was published in March 1971.

What, if anything, can be scientifically concluded about the deleterious effects of 2,4,5-T vis-à-vis TCDD on humans and other animals? As this question is addressed, it must be noted first of all that it is impossible to prove scientifically that a given chemical is without hazard. The small number of known cases in which human ingestion of 2,4,5-T led to clinical illness offer no information on the minimal dosage of the compound that is toxic to man. In animals, however, the toxicity of 2,4,5-T is similar to that of 2,4-D. Thus some information on 2,4-D is of interest.

In an extreme case, ingestion of a single dose of more than 6500 mg (in excess of 90 mg/kg) by a human in an attempted suicide was indeed fatal. But in contrast, a patient was given 18 intravenous doses of 2,4-D during 33 days without any apparent side-effects. Each of the last 12 doses in the series was 800 mg (approximately 15 mg/kg) or more, and the last was 2000 mg (about 37 mg/kg). In a study reported in 1973, five human male volunteers ingested a single dose of 5 mg/kg without incurring detectable clinical effects. The 2,4,5-T was rapidly excreted by the subjects. $LD_{50}s$[2]

---

[2] LD_{50} = dose that kills 50% of a population of test animals; "lethal dose for 50%."

for highly purified 2,4,5-T and 2,4-D, following oral ingestion by laboratory animals, are usually on the order of hundreds of milligrams of chemical per kilogram of body weight. In marked contrast, the LD_{50}s for TCDD ingested orally by laboratory animals vary from 0.0006 mg/kg for male guinea pigs to 0.115 mg/kg for rabbits (mixed sexes). As far as human hazard is concerned it seems certain that any danger of 2,4,5-T formulations resides in their TCDD content. While data are too limited for a firm conclusion, there is no evidence presently available that TCDD as a contaminant in 2,4,5-T is likely to be encountered by animal or man in sufficient dosage to cause toxic reaction. In its report, the U. S. President's Advisory Committee concluded that, "No evidence has been found of adverse effects on human reproduction in three separate locations, namely Vietnam; Globe, Arizona; and Sweden where pregnant women have allegedly been exposed to high levels of 2,4,5-T."

In concluding this subject two hypothetical examples may be of interest. First, a normal application (e.g., 2 lb/acre) of 2,4,5-T, for example, usually would constitute dispersal of no more than 450 μg of the dioxin on each treated acre, assuming a dioxin content of <0.5 ppm. Second, data from experimental animals indicate that immediately after spraying 2,4,5-T at accepted rates on forests, a 100-lb human would have to absorb all of the TCDD on 5 acres of forest to suffer a minimum detectable toxic effect. That would be the amount of TCDD suspected to be sufficient to cause birth defects in humans, if humans are as sensitive as laboratory animals.

Consideration of 2,4-D and other phenoxy-acid-type auxins calls to mind another important feature of those compounds, namely, the relationship between positions of halogen atoms on the ring and biological activity. For example, 2,4,5-T is a very potent auxin, whereas 2,4,6-T is inactive. In fact the latter compound competitively inhibits the action of the former and therefore is logically termed an "antiauxin." There are numerous other structure–activity relationships which could be mentioned, but the number of generalizations which hold true for all cases is few.

Natural Occurrence of Auxins

IAA is also synthesized in very many species of nonseed plants, including many bacteria, fungi, and algae, and may be ubiquitous in the plant kingdom. For many years, IAA was the one critically identified auxin, and it came to be thought of as the only natural auxin. However, there have been numerous reports of other substances in plants which show similar capacity to stimulate growth in auxin bioassays. Some of these substances are very obviously related to IAA, and, in the opinion of many, they simply represent alternative sources or precursors for IAA and are really active only upon conversion to IAA. At this time it must be said that some uncertainty still exists as to whether a number of indolic compounds may be active compounds, biosynthetic intermediates, degradation products, storage or transport forms, detoxification products, or artifacts. But, as K. V. Thimann wrote in 1969, ". . . in spite of frequent suggestions to the contrary, IAA remains the only naturally-occurring auxin whose existence is definitely established, and the wide occurrence as well of such a number of its derivatives makes it evident that IAA is very widely—perhaps universally—distributed throughout the plant kingdom."

Auxin Biosynthesis

For a compound to fulfill the role of a hormone, it should be apparent that its own concentration must be rigidly controlled. The control of concentration of auxin in the various parts of the plant is complex, involving multiple processes. One means of regulation of auxin levels is control of biosynthesis in situ. We shall now turn to a discussion of auxin biosynthesis and consider later other processes that contribute to regulation of auxin levels.

The amino acid tryptophan is commonly regarded as the precursor for the biosynthesis of auxin (IAA) in plants, and the biosynthetic pathways that have been worked out are illustrated in Fig. 2-8. By one pathway, tryptophan is converted to indolepyruvic acid via a transaminase reaction, which requires an α-keto acid and pyridoxal phosphate in addition to the enzyme. Indolepyruvic acid is next decarboxylated to indoleacetaldehyde in a reaction requiring a decarboxylase and thiamine pyrophosphate. An aldehyde dehydrogenase, for which nicotinamide adenine dinucleotide is the most effective coenzyme, then oxidizes indoleacetaldehyde to IAA.

The second major pathway involves an initial decarboxylation of tryptophan to form tryptamine. Catalysis by an amine oxidase next converts tryptamine to indoleacetaldehyde, which is, in turn, oxidized to IAA.

In some plant systems, one or the other of these pathways apparently occurs to the exclusion of the other. In other systems both pathways are operative.

Several other indolic compounds besides those already mentioned can function as precursors of IAA in particular higher plant systems. For example, cucumber (*Cucurbita pepo*) seedlings contain appreciable quantities of indole-3-ethanol (or tryptophol) which is readily converted to indoleacetaldehyde in that system. Members of the Cruciferae or Brassicaceae contain indole-3-acetonitrile and a nitrilase which converts the nitrile directly to IAA. Indoleacetamide occurs in some plants (e.g., tomato), which can reportedly go to IAA via a hydrolase reaction. Still other indolic compounds have been implicated as auxin precursors in particular higher plant systems.

Figure 2-8. Pathways of auxin biosynthesis. Enzymes: 1, tryptophan transaminase; 2, indolepyruvate decarboxylase; 3, indoleacetaldehyde dehydrogenase; 4, tryptophan decarboxylase; 5, amine oxidase; 6, indoleethanol oxidase; and 7, nitrilase.

An intriguing observation which has been reported many times is that D-tryptophan is often as active or even more active than L-tryptophan as an auxin precursor, even though the latter is the only isomer which occurs naturally in the proteins of higher plants. This enigma was answered by Miura and Mills (1971) for a system using cell cultures of tobacco (*Nicotiana tabacum* L.). They demonstrated the presence in the tissues of a tryptophan racemase which converts D-tryptophan to L-tryptophan in a reversible reaction. Since the pool size of free tryptophan in most higher plants is comparatively very small, it appears that the synthesis and utilization of tryptophan is subject to rigid control. Since D-tryptophan apparently can undergo transamination to form indolepyruvic acid and ultimately IAA, it seems possible that conversion of L-tryptophan to D-tryptophan prior to transamination may be important in controlling the levels of free tryptophan and of IAA. In other words, conversion of L-tryptophan to D-tryptophan may be a means of ensuring a pool of the amino acid for particular pathways including, but not restricted to, auxin biosynthesis.

There is another reaction for which D-tryptophan is a substrate in at least some plant tissues, and that is the formation of malonyl-D-tryptophan. M. H. Zenk suggested many years ago (1964) that the transformation of some unspecified indole auxin precursor to D-tryptophan, which is immobilized and accumulated as malonyl-D-tryptophan, might be significant in regulating the biosynthesis of IAA. This reinforces the notion that D- and L-tryptophan racemization may be important in the regulation of IAA biosynthesis.

It has been suggested by several investigators that the indole ring need not necessarily be derived from tryptophan, but might be synthesized directly in the course of auxin biosynthesis. In other words, the idea has been expressed that auxin biosynthesis may occur by a tryptophan-independent pathway. For example, Winter (1966) suggested, on the basis of very limited studies with *Avena* coleoptile sections, that tryptophan is not a precursor of IAA in that system. He proposed a hypothetical pathway according to which anthranilic acid is converted to indole, and that indole by unknown reactions could, in turn, be converted to tryptamine and on to IAA. However, Winter's scheme lacks proof. In the absence of convincing experimental evidence to the contrary, we shall conclude that tryptophan is the precursor of IAA.

A serious potential problem confronting investigations of auxin biosynthesis is microbial contamination, since numerous species of microorganisms are known to convert tryptophan to IAA. The influence of epiphytic bacteria on auxin metabolism in preparations from pea plants has been documented extensively in recent years by Libbert et al. in Germany. Hence several investigators have performed experiments to determine the possible influence of microbial metabolism in the system they have employed. Utilizing a variety of techniques—e.g., involving aseptic culture of seedlings and antibiotics—it has been concluded that in enzyme extracts of pea shoot tips and other higher plant tissues, tryptophan is indeed converted to IAA by enzymes present in the tissues.

At the present state of our knowledge, the possibilities must be considered that: (1) there may be, and probably are in particular cases, different pathways of IAA biosynthesis from tryptophan in different species; (2) that more than one pathway may be operative in a single species; and (3) that these pathways may differ in different parts of the same plant or at different stages of growth in the same plant part.

"Free" and "Bound" Auxin

The natural auxin indoleacetic acid (IAA) exists in a variety of chemical states in plant tissues. Free auxin is considered to be that auxin which is readily extractable (e.g., 2 hours in diethyl ether at 0°C in the dark) while, traditionally, bound auxin is considered to be that auxin liberated from tissues when subjected to enzymolysis, hydrolysis, or autolysis. Free auxin, IAA per se, apparently is the form immediately utilizable in growth. Bound auxin apparently can exist in various forms, and the forms vary among different species (Fig. 2-9). Bound auxins generally are con-

Figure 2-9. Some forms of bound auxin.

sidered storage forms from which IAA can be released or detoxification products which form in the presence of relatively high levels of IAA. Auxin glycosyl esters are especially abundant in seeds and storage organs and apparently represent inactive, storage forms of hormone from which IAA can be enzymically released. Ascorbigen and glucobrassicin may also be considered bound forms of auxin in some species of the Cruciferae or Brassicaceae. Auxin peptides, e.g., indoleacetylaspartic acid, are generally considered detoxification products which are irreversibly formed and protect tissues from the accumulation of excessive auxin. Auxin–protein complexes are hypothetical associations, presumably by weak hydrogen bonds, between IAA and specific proteins.

Some synthetic auxins also form bound auxin. For example, the following auxins (besides IAA) reportedly can conjugate with aspartic acid in pea epicotyl sections: indolepropionic, indolebutyric acid, benzoic acid, and 2,4-D. Actually, very little 2,4-D was destroyed or conjugated. This auxin is relatively very persistent and quite resistant to enzymic degradation and conjugation. Naphthaleneacetic acid (NAA) was reported to form a glycosylester in plants from several families, as well as to form a peptide with aspartate.

Destruction of IAA

As already emphasized, the level of free auxin in tissues must be rigidly controlled. A third process by which auxin levels are regulated, besides regulated synthesis in situ and reversible and irreversible formation of bound auxin, is degradation to inactive compounds. Auxin destruction is by two basic oxidative processes: (1) enzymatic reaction; and (2) photooxidation.

The most important oxidative process physiologically is enzymatic destruction, catalyzed by an enzyme known as IAA oxidase. An accurate description of IAA oxidase is that it is a peroxidase acting as an oxidase (Fig. 2-10). The enzyme is at least partially independent of exogenous H_2O_2, when assayed in vitro, apparently because it carries a peroxide-producing system along with it either as an impurity or as a bound flavin moiety. Generally, in assays in vitro, only about 1/10 mole of exogenous H_2O_2 is needed per mole of IAA oxidized, and peroxidation in the strict sense, that is, by H_2O_2 in the absence of O_2, does not occur. One mole of O_2 is taken up, and 1 mole of CO_2 evolved from the carboxyl group. One of the dominant products is 3-methylene-oxindole. Another seen sometimes is indolealdehyde. The enzyme system from higher plants requires Mn^{2+}.

Since P. M. Ray had shown by 1960 that IAA oxidase appeared to be a peroxidase function, R. L. Hinman and J. Lang set out in 1965 to investigate IAA oxidation in the presence of purified horseradish peroxidase in the absence of exogenous H_2O_2. They recognized that the IAA-oxidase systems are of the peroxidase type (Ray 1958, 1960) and considered that the natural IAA oxidases and purified peroxidases would probably yield closely related if not identical products. The first reaction in their scheme (Fig. 2-11) is a one-electron oxidation of the indole ring in which Fe^{3+}, NO_2^-, and peroxidase all function in the same way to produce a free radical. Two major enzymic pathways were demonstrated overall. One leads to the formation of 3-methyleneoxindole and the other to 3-indolealdehyde. The formation of the latter product is favored by high auxin concen-

Methylene oxindole

Figure 2-10. Summary of the IAA oxidase reaction.

Figure 2-11. Oxidative metabolism of IAA catalyzed by peroxidase and IAA oxidase. (Redrawn with modifications, by permission, from: Hinman, R. L. and J. Lang. 1965. Peroxidase-catalyzed oxidation of indole-3-acetic acid. *Biochemistry* **4**: 144–158. Copyright by the American Chemical Society.)

trations and the presence of cytochromes and cytochrome oxidase. To a lesser extent a pathway leading to nonenzymic opening of the indole ring of the hydroperoxide was demonstrated. IAA oxidase reaction is accelerated by monophenols (e.g., 2,4-dichlorophenol); it is inhibited by *ortho*-diphenols such as catechol and pyrogallol. Flavanols act in the same way; that is, kaempferol promotes and quercetin inhibits IAA oxidase.

The photooxidation of IAA is a very different reaction from the enzymatic one and appears to have little physiological significance. Riboflavin, eosin, and other fluorescent dyes are effective catalysts in assays in vitro. One mole of O_2 is taken up per mole of IAA destroyed, as in the enzymic, IAA oxidase reaction, and the products apparently are identical, including 3-methyleneoxindole and 3-indolealdehyde. Relatively very large light dosages are required, and the process seems to have no bearing on phototropism or other physiological phenomena.

V. Tuli and H. S. Moyed (1967) showed directly that extracts of pea seedlings (*Pisum sativum* var. Alaska) oxidized IAA to 3-hydroxymethyloxindole, and at physiological pH this compound was readily dehydrated to 3-methyleneoxindole. The extracts of pea seedlings also contained a reduced triphosphopyridine nucleotide-linked enzyme which reduced 3-methyleneoxindole to 3-methyloxindole. 3-Methyleneoxindole

is known to be a potent bacteriostatic agent. It is a potent sulfhydryl re-
agent, which reacts rapidly with substances such as glutathione and CoA,
and is a potent inhibitor of SH-containing enzymes. 3-Methyloxindole is
nontoxic. The observed enzymic reactions were also seen to take place in
intact pea seedlings.

It has been reported that the peroxidase of peas exists as a number of
isozymes. Isozyme patterns apparently change during ontogeny and may
vary among the organs of a single plant.

Addition of IAA to green tissues enhances peroxidase activity, but this
does not appear to be true enzyme induction of IAA oxidase. While treatment of plant tissues with auxins (IAA or other) often increases the peroxidase activity, auxins which are not oxidized act as well as IAA. Hence
the change is evidently not a true induction enzyme activity but a result of
generally enhanced synthesis.

It was noted previously that monophenols and *ortho*-diphenols modify
IAA oxidase activity, and there is evidence that some endogenous phenols affect enzymic activity in a significant way. For example, pea plants
contain the flavonols kaempferol and quercetin as the triglucosides and
coumaroyl (coumaric acid) triglucosides. When an etiolated pea plant is
irradiated with white or red light, interesting correlative effects of red
light occur with respect to the flavonols. Red light promotes apical bud
growth and inhibits subapical stem elongation. In a reportedly correlative
manner, red light stimulates the synthesis of kaempferol, a cofactor of
IAA oxidase, in stems, and stimulates synthesis of quercetin, an inhibitor
of the enzyme, in leaves. Since phenolic substances occur widely in
plants, they must be assumed to have some importance as modifiers of
auxin destruction.

Auxin Transport

For many years, actually since the classical experiments of F. W. Went
with *Avena* coleoptile sections in 1932, polarity of auxin transport has
been recognized. Apical dominance and other growth correlations (e.g.,
cambial activation) in the plant are correlated with polar movement of endogenous auxin. Tropisms are correlated with asymmetrical movement of
endogenous auxin. Indeed, the influences of auxin on growth and development are, in general, reflections of the polar pattern of its distribution.

The movement of endogenous auxin in shoots is predominately basipetally polar from morphological apex to morphological base (Fig. 2-12).
This basipetal polar movement occurs not only in segments of coleoptiles
or Coleus stems lacking vascular tissue but also in veins of leaves. Thus,
both parenchymatous tissue and vascular tissue appear to be capable of

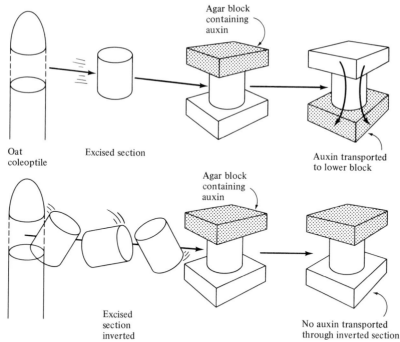

Agar block
containing
auxin

Oat
coleoptile   Excised section

Auxin transported
to lower block

Agar block
containing
auxin

Excised
section
inverted

No auxin transported
through inverted section

Figure 2-12. Early experiment of F. W. Went, demonstrating basipetal polar transport of endogenous auxin in excised sections of *Avena* coleoptile. (Redrawn, with permission, from Bonner and Galston, 1952.)

polar transport. There actually is remarkably little direct evidence on the question of how endogenous auxin moves in the intact whole plant; there are many indications, however, that endogenous auxin moves readily in the phloem. And indirect evidence suggests that polar basipetal movement may occur in the cambia of trees. Polar basipetal movement diminishes in older parts of stems, coleoptiles, and petioles. Acropetal transport in at least two systems from shoots, oat coleoptiles and bean petioles, apparently is nonmetabolic and appears to be diffusion.

In excised coleoptile and stem sections, IAA is not the only substance which can move in the polar transport system. Polarity of movement has also been demonstrated for several synthetic auxins, including indolebutyric acid, naphthaleneacetic acid, and 2,4-D. In general, growth regulators (including IAA, 2,4-D, etc.) applied to intact roots or shoots move in both xylem and phloem, but this movement is nonpolar, and velocities of 50 mm/hour or more are not uncommon.

Auxin transport in roots is less completely understood than movement in shoots. The most convincing data, from experiments in which movement of [^{14}C]IAA in root segments has been studied, indicates that movement is acropetal, that is, from the base of the stem toward the root tip.

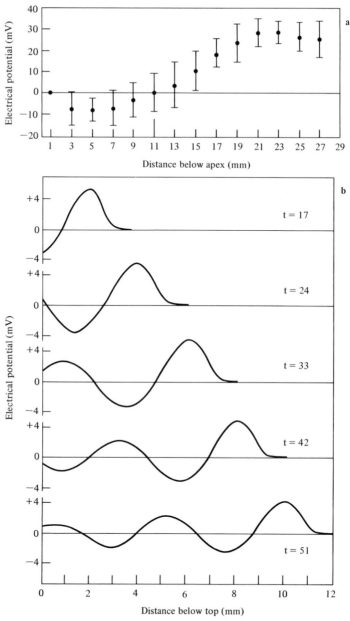

Figure 2-14. Gradients in electrical potential in *Avena* coleoptiles. **a,** Electrical potential plotted against distance from the apex of an intact coleoptile; **b,** wave of electrical disturbance which accompanies polar movement of exogenous IAA down a decapitated coleoptile. In **b** the potential plotted is the departure of the potential from its average value. The average potential distribution is illustrated at each of five times (*t*, in minutes) after the application of IAA. (Redrawn, with permission, from Newman, 1963.)

of 100 cells were to come out of the basal end of each cell, over 10,000 times as much auxin would be found at the basal end of the tissue than at the acropetal end.

A very intriguing question is: What is the source of acropetally moving auxin in the roots of intact plants? It is unlikely that such auxin is merely from the downward moving stream of auxin that the shoot produces, because—judging by results with sections—there is a graded decrease in the amount of auxin transported down the bean hypocotyl to the point where little or none moves through sections from the stem–root transition region. There is evidence that auxin is synthesized in root tips, but it is difficult to envisage how auxin from the root tip comes to be transported acropetally in subapical sections of the root. The point needs investigating, particularly with a critical appraisal of how accurately the results with sections apply to the intact plants.

Relationships between Auxin Content and Growth

Multiple factors determine the amounts of endogenous auxin in a particular part of a seed plant at a given time. Gain by import from primary centers of auxin synthesis and loss by export via polar transport is one important factor, which has just been discussed. Auxin is synthesized in relatively large amounts in only a few localized centers, but it is transported through all the living tissues of the plant. Shoot tips, including particularly the young leaves, are the center of most abundant auxin synthesis in the vegetative seed plant. Other rich sources are enlarging leaves, flowers, fruits, and seeds. The enzymes for the conversion of tryptophan to IAA occur generally throughout the plant and are especially active in regions of intense metabolic activity, such as meristems, expanding leaves, fruits, and root tips. Other factors, already discussed, are: synthesis in situ, enzymic destruction, and reversible and irreversible formation of bound auxin and release of free auxin.

In general, a reasonably good correlation is found between relative auxin content and relative growth in the various organs of seed plants, particularly taking into account the marked differences in sensitivity of the different parts of a plant to auxin (Fig. 2-15). And, there are available data to show that auxin content is indeed influenced by the various factors which have been previously discussed. For example, Figs. 2-16, 2-17, and 2-18 illustrate the amounts of auxin (presumably free auxin, specifically, as judged by the methods used) in different parts of a monocot and two species of dicot seedling, respectively.

IAA oxidase preparations from different parts of an angiosperm seedling tend to reveal, by in vitro assay, an inverse correlation between en-

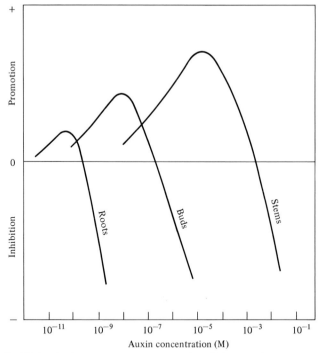

Figure 2-15. Biphasic dose–response curves for different plant organs and IAA. (Redrawn, with permission, from Leopold, 1955.)

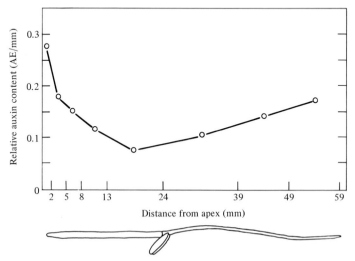

Figure 2-16. Distribution of endogenous auxin in an etiolated *Avena* seedling. The auxin content is expressed in *Avena* Einheits (AE), units no longer in common usage. (Redrawn, with permission, from data of Thimann, 1934, in Leopold, 1955.)

Figure 2-17. Comparative diffusible auxin content of different parts of a seedling of broad bean (*Vicia fava*). The numbers refer to *Avena* Einheits of diffusible auxin per hour. (Redrawn, with permission, from data of Thimann and Skoog, 1934, in Leopold, 1955.)

Auxin from leaves

Auxin from buds

2.2

1.5

12.0

0.5

0.4

0.6

0.6

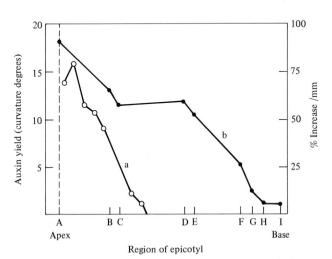

Region of epicotyl

Figure 2-18. Relative growth rate, **a**, and distribution of diffusible endogenous auxin, **b**, in the epicotyl of a 9-day-old light-grown pea seedling. (Redrawn, with permission, from Scott and Briggs, 1960.)

zyme activity and auxin content. That is, IAA oxidase activity tends to be comparatively low in regions of relatively high auxin content (Fig. 2-19).

In 1969 investigations were conducted on the comparative net biosynthesis of IAA from tryptophan in cell-free enzyme extracts of different parts of pea (*Pisum sativum* L. cv. Alaska) seedlings and immature pea seeds (Table 2-1). Tryptophan was converted to IAA in enzyme extracts of all parts of the light-grown pea seedlings and in extracts of immature pea seeds. Auxin yields are expressed as the amount of radioactivity from DL-[2-^{14}C]tryptophan incorporated into [^{14}C]IAA per gram fresh weight of tissue. The unit fresh weight is considered a physiologically more meaningful parameter than unit protein or nitrogen for comparing the auxin-synthesizing capacities of the different parts of seedlings, because of the marked differences in protein and nitrogen concentration of the plant materials. On this basis, the regions of most active auxin production apparently were the terminal bud, young stem below the terminal bud, and young leaves. Lesser amounts of net auxin biosynthesis occurred in enzyme extracts prepared from older stems and leaves and from root tips. Thus, in general, the data agree well, to the extent that they are comparable, with the amounts of "free" auxin obtained by diffusion and extraction from young light-grown pea seedlings. Developing pea seeds exhib-

Specific activities
(μg IAA destroyed $\cdot$ mg protein^{-1} $\cdot$ hour^{-1})

230
614
285
1410
1740
1518
2460
2375
3015

Figure 2-19. Relative magnitude of IAA oxidase activity (specific activities) in different parts of an etiolated pea epicotyl. (Redrawn, with permission, from Galston, 1967.)

Table 2-1. Comparative Net Biosynthesis of [¹⁴C]IAA in Cell-Free Extracts of Different Parts of Pea Plants[a,b]

Plant part(s)	dpm incorporated into [¹⁴C]IAA/g fresh wt/2 hours	Mean fresh wt of parts/plant (mg)	mg N/g fresh wt
Terminal bud	61,166	24	6.31
Stem above 5th node	40,181	98	1.59
Leaves above 5th node[c]	51,776	77	4.96
Shoot tip (all above 5th node)	38,933	305	3.55
Stem below 5th node	19,200	478	0.68
Leaves below 5th node[c]	10,240	545	1.62
Apical 2 cm of primary root	7,786	34	0.98
Developing seeds[d]	56,000	171	2.22
Developing seeds[e]	89,266	—	—

[a] From Moore (1969) with permission.
[b] Collective data from three separate experiments.
[c] Including stipules, rachis, leaflets, and tendrils.
[d] Approximately half-grown "Alaska" pea seeds.
[e] Approximately half-grown "Telephone" pea seeds.

ited an auxin-synthesizing capacity which was comparable to or higher than that of terminal buds of seedlings.

Two considerations ought to be borne in mind in interpreting the data in Table 2-1, however. In the first place, it is not known to what extent the auxin-synthesizing capacities in cell-free enzyme extracts reflect the actual in vivo capacities of the various plant parts to produce IAA from tryptophan. Second, in these experiments net IAA production was measured, and there was no accounting for concomitant auxin degradation by the IAA oxidase system in the preparations. Previously it was shown that [1-¹⁴C]IAA (carboxyl-labeled), when added to cell-free enzyme extracts of shoot tips in estimated substrate quantities, was enzymically decarboxylated. Hence it is reasonable to assume that IAA oxidase activity was present, and at variable levels, in the enzyme extracts of the various plant parts.

In summary there appears to be a good correlation between capacity for net auxin biosynthesis from tryptophan and endogenous free auxin content of the various parts of young green pea seedlings. There still remains the question of whether there are differences in the capacities of the various parts of the seedling plant for gross auxin production, or whether the differences in apparent auxin-producing capacity noted in the present investigation are correlated primarily with differential IAA oxidase activity. Scott and Briggs, while also considering other possible contributing factors, felt in 1963 that auxin destruction seemed to be the best alternative to account for the variations in "free" auxin content in different parts of green pea epicotyls. Based on all available data it appears probable that the different parts of the pea seedling do differ in potential auxin-synthesizing capacity as well as in IAA oxidase activity, and that differential auxin biosynthesis in situ is one factor among many which accounts for variations in auxin content in the various parts of pea seedlings.

Correlative Differences in Auxin Relations between Etiolated and Light-Grown and Dwarf and Normal Plants

There is considerable interest in the possible correlative differences in auxin relations between etiolated and light-grown seedlings of the same species and cultivar and between plants of the dwarf genotype and those of the normal or tall genotype of different cultivars of the same species. *Pisum sativum,* garden pea, is an example of an herbaceous annual species which has been used quite extensively in research in this area, and the hormone field generally. Both genetic dwarf and normal (tall) varieties of pea are available, and the genetics of several varieties is fairly well known. Dwarf varieties mature at heights of 30 cm or less, whereas some tall varieties exceed a meter in shoot height at maturity. Because of their self-fertilization and high degree of homozygosity, each variety is phenotypically very uniform.

Regarding etiolated and light-grown seedlings of the same cultivar of *Pisum sativum,* there is a marked difference in sensitivity of excised stem sections, with the former being about two orders of magnitude more sensitive than the latter (see Fig. 6-9). Other interesting physiological differences between the etiolated and light-grown specimens are known, and many of these were described by T. K. Scott and W. R. Briggs in 1963, using seedlings of the Alaska variety, a normal, early-flowering type (Table 2-2).

The hormonal basis for dwarfism presents some very interesting questions, which have been answered for only a few species. In the case of peas, etiolated specimens of both dwarf and normal genotype exhibit nearly identical rates of stem elongation (Fig. 2-20). That is, dwarfism becomes readily apparent only in the light-grown condition.

T. C. Moore in 1967 compared the net biosynthesis of IAA from tryptophan in cell-free enzyme extracts prepared from the shoot tips of etiolated and light-grown seedlings of the Dwarf and Tall Telephone cultivars. This comparison of maximum net IAA biosynthesis revealed an order, on a unit protein basis, of: light-grown tall > light-grown dwarf > etiolated tall $\simeq$ etiolated dwarf.

It was concluded in that study that the different rates of stem elongation among etiolated and light-grown dwarf and tall pea seedlings are correlated with small differences in auxin relations. However, it was also

Table 2-2. Comparison of Light-Grown and Dark-Grown Pea Seedlings with Respect to Structure, Growth Patterns, and Growth Substance Relationships[a]

	Light-grown	Dark-grown
Morphology and anatomy	Five internodes, top two, true stem; basal three, transition zone	Three internodes, all transition zone
Growing region	Top internode, evenly distributed	Top internode, primarily apical region
Extractable auxin	High, no change from apex to base of top two internodes, decrease basally through transition zone	Decrease, apex to base
Diffusible auxin	Growing region, decrease, apex to base; no change through 2nd internode; decrease, apex to base of transition zone	Not obtainable
Auxin transport	Polar, 10–12 mm/hour, saturated with endogenous auxin only in transition zone	Possibly incompletely polar, apparently 6–7 mm/hour
IAA oxidase activity	At least partially masked by inhibitor	High, increasing, apex to base
IAA oxidase inhibitor activity	High, decreasing, apex to base	Low
Response, sections from growing zone, to IAA	Optimal at high concentrations	Optimal at low concentrations
Response, sections from growing zone, to gibberellic acid	Optimal at high concentrations	Optimal at low concentrations

[a] Adapted from Scott and Briggs (1963) with permission.

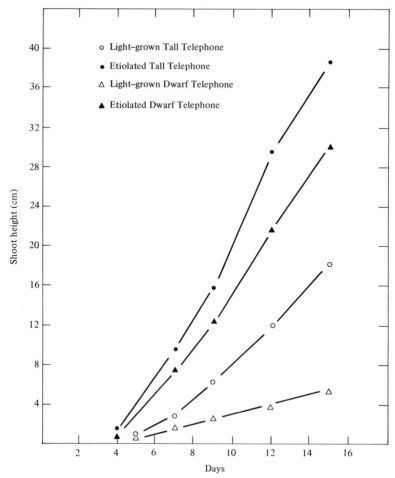

Figure 2-20. Growth curves of etiolated (closed symbols) and light-grown seedlings (open symbols) of Tall (circles) and Dwarf (triangles) Telephone peas grown in growth chambers. Each point represents the mean value for 20 plants. (Unpublished data of Moore, 1964.)

concluded that correlative differences in endogenous gibberellin relations probably are of much greater importance. The hormonal basis of dwarfism is discussed more comprehensively in Chapter 3.

Mechanism of Auxin Action

Some Fundamental Terms and Concepts

As L. G. Paleg stated in 1965, there is merit in applying different connotations to two terms, "mechanism" and "mode" of action, which often are used synonymously. When a hormone acts upon a responsive plant system, it of course enters into some direct and specific molecular interaction which results, eventually, in a manifestation of measurable effect (biochemical or physiological response, e.g., cell elongation). But there really are two aspects of the hormone action involved: (1) the direct and specific molecular interaction; and (2) the succeeding series of steps (events, processes) which result in the measurable biochemical or physiological response. The former is the "mechanism" of action; the latter is the "mode" of action.

By definition, therefore, every hormone has its own distinctive mechanism(s) of action, even though the manifestation of the hormonal mechanism may depend upon the prior, concomitant, or subsequent role of another factor(s). It follows that in one system a hormonal mechanism may lead to one physiological manifestation, while in a second system the same mechanism may lead to a completely different manifestation. This difference can be brought about because the second system has more (or less) of another hormone, is from a different tissue or organ, or other causes. Thus, it is to be emphasized that auxins have their own distinctive mechanisms of action, though their modes of action frequently may be responsive to other factors.

Preliminary Considerations Concerning the Mechanism of Auxin Action

Ever since the discovery of each type of plant hormone, great effort has been made to attain a biochemical understanding of the manner by which each type exerts its diverse effects. A major difficulty in elucidating the mechanism of auxin (and other hormone) action is the multitude of different kinds of physiological processes auxin controls. One might naturally wonder whether there is only one fundamental site and mechanism of auxin action, which leads to diverse physiological responses, or whether there are multiple sites and mechanisms of action. The truth is that we do not yet know. However, very significant progress toward ultimate elucidation of the mechanism of auxin action has been made in recent years, particularly since circa 1960. Let us tabulate some of the most important known facts about auxin action that any theory (model) of auxin action must accommodate:

1. All auxin-induced physiological responses require constant presence of the hormone. In the case of auxin-induced (IAA) cell elongation in coleoptiles, an increased growth rate is evident within 10 minutes or less after presentation of the auxin; upon withdrawal of IAA a decline in growth rate is evident within 10 minutes and growth rate is back to control level within about 40 minutes after withdrawal of the auxin.

2. Auxin (like hormones generally) is active in extremely small concentrations (e.g., 10^{-6} M). Therefore, some form of amplification of an initial triggering response must occur; e.g., amplification might occur by

the hormone: (a) acting as an allosteric effector activating certain enzymes; (b) stimulating the synthesis of certain enzymes; or (c) evoking
a change in membrane permeability.

3. It is clear that continued RNA (specifically mRNA) and protein synthesis are essential for cell elongation responses to be sustained for as
long as several hours. The ability of auxin to enhance the rate of cell
elongation is dependent upon new RNA (mRNA) and protein
synthesis.

4. It is equally clear that some auxin responses apparently occur too rapidly to involve gene activation as the primary action of auxin. For examples: elongation in coleoptile and stem sections occurs in < 10 minutes; increase in protoplasmic streaming can often be observed within
20 seconds to a few minutes; and auxin alteration of membrane potentials occurs in 10 to 15 minutes or less.

Water Relations in Auxin-Induced Cell Enlargement

The auxin function which has been investigated most extensively is auxin-
induced cell enlargement. Auxin is required specifically for cell wall
loosening but not for cell enlargement itself. Reviewing the equation relating the major parameters of cell water potential, we see that:

$$\Psi \text{ (water potential)} = \Psi_\pi \text{ (osmotic potential)} + \Psi_p \text{ (pressure potential)}$$

For water uptake to be promoted by auxin, the water potential inside
the cell must become more negative than outside. Conceivably, Ψ could
become more negative either because:

1. the osmotic potential (Ψ_π) becomes more negative, or
2. the pressure potential (Ψ_p) ($\equiv$ turgor pressure) decreases.

What actually happens in the auxin response are the following sequential events:

1. Auxin increases wall loosening, thus decreasing the resistance of the
wall to stretching and allowing the wall to yield to pressure.
2. The pressure potential (Ψ_p) thus decreases.
3. More water osmoses in because of the more negative internal Ψ.
4. The cell volume is thereby increased and the wall is irreversibly
stretched.

The cell wall loosening action of auxin really is a dual one—an initial
effect on elasticity and a succeeding effect on plasticity (Fig. 2-21). Of
these two effects, the latter is most important, for cell enlargement is most
directly correlated with the effect on cell wall plasticity (Fig. 2-22). Upon
completing an increment of enlargement due to one exposure to a growth-

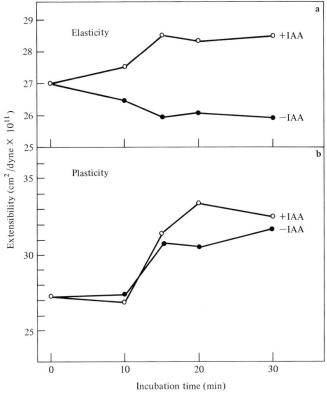

Figure 2-21. Effects of auxin on cell wall elasticity, **a**, and plasticity, **b**, in oat co-leoptiles. (Redrawn, with permission, from Masuda, 1969.)

promoting concentration of auxin and withdrawal of the hormone, the rigidity of the primary cell wall reverts to its original state.

The Nature of Auxin-Induced Cell Wall Loosening

The primary cell wall consists of a framework of cellulose microfibrils which are cross-linked by other polysaccharides. Peter Albersheim and associates of the University of Colorado have worked out the most extensive analysis of the macromolecular components of primary cell walls, using cultured sycamore cells (Table 2-3) and also have created a model of the molecular architecture of the wall (Fig. 2-23).

Albersheim and associates have also worked out a model for auxin-induced cell expansion, which includes the following features. Since walls extend throughout their length, for a wall to be extended, it is necessary that the cellulose fibrils be able to slide along their length relative to each other. Also, walls that have extended are essentially as strong as those that have not, indicating that the number of cross-links and the amount of

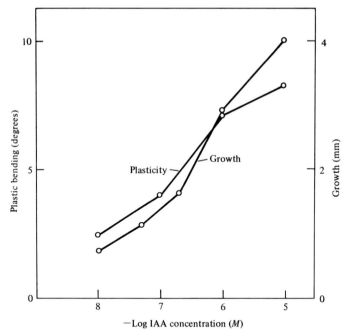

Figure 2-22. Direct correlation between auxin effects on growth and cell wall plasticity. (Redrawn with modifications, by permission, from Bonner, 1961.)

material in an extended cell wall are insignificantly different from those in an equivalent area of cell wall prior to expansion. In other words, if a wall extends as a result of cleavage of cross-linking bonds between polymers, a mechanism must exist for the formation of new cross-links at the termination of an increment of cell expansion.

Examination of their model of the cell wall led Albersheim and associ-

Table 2-3. Macromolecular Components of the Walls of Suspension-Cultured Sycamore Cells[a]

Wall component	Cell walls (%)
Arabinan	10
3,6-Linked arabinogalactan	2
4-Linked galactan	8
Cellulose	23
Protein	10
Rhamnogalacturonan	16
Tetra-arabinosides (attached to hydroxyproline)	9
Xyloglucan	21
Total	99

[a] With permission, from Talmadge et al. (1973).

— cellulose elementary fibril

– xyloglucan

— wall protein with arabinosyl tetrasaccharides glycosidically attached to the hydroxyproline residues

— total pectic polysaccharide

— rhamnogalacturonan main chain of the pectic polysaccharide

– arabinan and 4-linked galactan side chains of the pectic polymer

— 3,6-linked arabinogalactan attached to serine of the wall protein

— unsubstituted seryl residues of the wall protein

Figure 2-23. Tentative model of the molecular architecture of sycamore cell walls. (With permission, from Keegstra et al., 1973.)

ates to formulate a specific hypothesis regarding the breaking and reforming of cell wall cross-linkages. They found that the only noncovalent cross-links among the structural polymers of the cell wall evidently are the hydrogen bonds between the xyloglucan polymers and the cellulose microfibrils. Movement of the xyloglucan chains along the cellulose microfibrils could be accomplished by a mechanism—enzymic or nonenzymic—which catalyzes breakage and reformation of the hydrogen bonds, allowing the xyloglucan to creep inchworm-fashion along the cellulose microfibrils. The rate by which the xyloglucan polymers move along the cellulose microfibrils should be increased by conditions, such as high hydrogen-ion concentrations, which weaken hydrogen bonds. An enhanced rate of creep would be completely and immediately reversed by raising the pH or lowering the temperature. Furthermore, one can readily conceive of unidirectional creep in walls under tension since the xyloglucan polymers are covalently linked at their reducing ends to noncellulosic wall components. Therefore, when neighboring cellulose fibrils are pushed in an opposite direction by hydrostatic pressure, the xyloglucan polymers on each fibril would creep in only one direction relative to the cellulose fibril to which they are attached.

The further intriguing suggestion was made by Albersheim and associates that auxin may activate a hydrogen-ion pump in the plasmalemma, hence lower the pH of the wall, enhance xyloglucan creep, and thereby increase cell wall loosening. We will return to this idea later in this chapter.

Sites of Auxin Action

Currently there are two major concepts as to the site of action of auxin. One of these centers on the cell wall as the site of action, a concept already introduced in considering the investigations by Peter Albersheim and associates. The other concept focuses on nucleic acid metabolism. Many cell-wall loosening phenomena may be related to auxin treatment in a manner independent directly of nuclear activities. Yet an enhancement of RNA synthesis is consistently associated with sustained growth responses to the hormone. Hence, we have two concepts which at first seem to be mutually exclusive and to indicate that there are multiple sites (and perhaps mechanisms) of auxin action. However, after examining the evidence on which each concept is based, we shall describe two hypothetical models which attempt to reconcile the two concepts into one. For now we shall first examine the evidence that auxin exerts its hormonal action by an effect on nucleic acid and protein metabolism. Then we shall discuss so-called rapid responses to auxin which do not depend upon a direct effect of the hormone on nucleic acid or protein metabolism.

Effects of Auxins on Nucleic Acid and Protein Metabolism

Abundant evidence indicates quite conclusively, and not at all surprisingly of course, that continued synthesis of RNA and protein is required for continued cell enlargement. Most investigations in recent years have utilized excised plant parts (sections of coleoptiles, stems, tubers), have measured rates of growth and of incorporation of radioactive precursors into RNA and protein, and have relied heavily upon the use of various inhibitors of RNA and protein synthesis.

Before getting into a detailed description of those types of investigations it will be well to provide a general characterization of effects of auxins on intact plants and excised tissues. For one thing, very few species of plants exhibit positive growth responses to applied auxin; cucumber (*Cucurbita pepo* cv. National Pickling) is one exception. Generally, intact plants respond to the auxin-type growth regulators by an inhibition of growth (both cell division and cell enlargement) and inhibition of nucleic acid synthesis in the normal growing points; the more mature stem tissues are activated relative to nucleic acid synthesis and massive cell proliferation. Treatment of intact seedlings with 2,4-D or some other auxin, while causing an increase in DNA and protein as well as in RNA, causes an apparent overproduction of RNA, resulting in increased RNA/DNA and RNA/protein ratios. The RNA accumulates mainly as ribosomal RNA. Thus, incidentally, the implications are that the herbicidal action of relatively high dosages of 2,4-D-type regulators relates to their hormonal effects, that is, evocation of massive RNA and protein synthesis rather than to some direct inhibitory action.

Excised tissues exhibit variable characteristics, depending upon source and age. When plant tissues are excised and cultured for a few hours in solution, there is usually a decrease in the RNA. In the case of many excised tissues, there may be no overall increase in protein during elongation.

J. L. Key and associates have done extensive work on the effects of auxin on nucleic acid and protein metabolism in elongating and maturing sections of hypocotyls of soybean (*Glycine max*) seedlings. Much of the following discussion is based on their studies. In their work, auxin brings about a general enhancement of the synthesis of all species of RNA, with the major increase in rRNA. However, since rRNA and tRNA have a longer mean life than mRNA, it is to be expected that they would be the species that would accumulate in response to auxin over long labeling periods. In excised elongating soybean hypocotyl, the RNA decreases during incubation. Growth-promoting concentrations of auxin cause the maintenance of RNA at or near the initial level. The enhanced incorporation of ^{14}C-labeled precursor into RNA in response to auxin indicates that auxin maintains the higher RNA level by enhancing synthesis. Growth-inhibiting concentrations of auxin inhibit ^{14}C-labeled precursor incorporation without appreciably affecting the RNA content relative to the control sections. In excised maturing soybean hypocotyl, auxin causes a large net synthesis of RNA, while the RNA content of the control tissues does not change. The net accumulation of RNA in response to auxin was associated with about a twofold enhancement of incorporation of [14]C- or ^{32}P-labeled precursor, and was linear after an initial lag of 2 to 3 hours. A definite but small enhancement of incorporation of ^{14}C-labeled precursor occurred during this lag period.

Incidentally, it should be noted that in many, if not most, investigations of auxin action, 2,4-D or some other synthetic auxin is used rather than the natural auxin IAA. This is because synthetic auxins such as 2,4-D are chemically more stable than IAA, more resistant also than IAA to IAA oxidase, and less susceptible than IAA to conjugation with particular amino acids. Yet the available evidence indicates that the several synthetic auxins which are commonly used behave in a manner identical to the hormonal action of IAA.

The evidence for a requirement for continued RNA and protein synthesis for continued cell elongation is derived from studies of inhibitors of RNA and protein synthesis, or of gene transcription and translation. Actinomycin D, an inhibitor of DNA-dependent RNA synthesis, inhibits both growth of elongating hypocotyl sections (Fig. 2-24) and incorporation of radioactive precursor into total RNA (Fig. 2-25). There is a lag period of 1 to 2 hours after presentation of auxin and inhibitor before inhibition is evident. Furthermore, much of the observed RNA synthesis apparently is not directly essential to growth, since low concentrations of actinomycin D inhibited RNA synthesis by as much as 30% without inhibiting growth (Fig. 2-26).

To determine the nature of the RNA required for cell elongation, Key et al. used the more specific inhibitor 5-fluorouracil, which selectively inhibits the synthesis of all types of RNA except messenger RNA (mRNA). 5-Fluorouracil inhibited total RNA synthesis up to 70% without affecting elongation of the hypocotyl sections (Fig. 2-27). These data indicate that the RNA required most immediately or directly to support protein synthesis and cell elongation over a period of several hours is mRNA.

When the relationship between the auxin-induced growth response and protein synthesis was examined, it was readily seen that protein synthesis, measured as incorporation of [^{14}C]leucine, was inhibited as predicted when growth was inhibited by actinomycin D (Fig. 2-28). When elongating hypocotyl sections were treated with cycloheximide, an inhibitor of cytoplasmic protein synthesis, a parallel inhibition of growth and protein synthesis also was observed (Fig. 2-29). These data show a close association between RNA and protein synthesis and the regulation of cell elongation by auxin. Moreover, in the experiments by J. L. Key and associates,

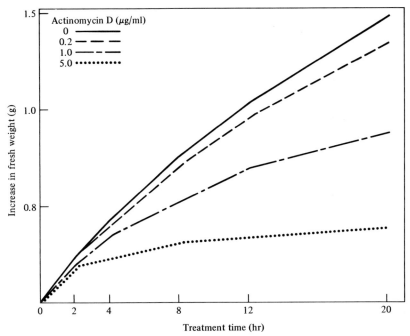

Figure 2-24. Inhibition of auxin-induced growth in elongating hypocotyl sections by actinomycin D. (Redrawn, with permission, from Key et al., 1967.)

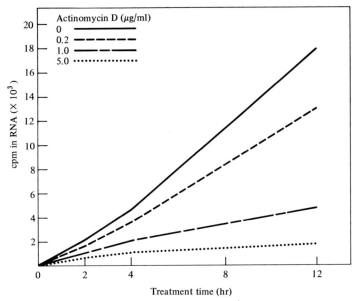

Figure 2-25. Inhibition of RNA synthesis in soybean hypocotyl sections by actinomycin D. (Redrawn, with permission, from Key et al., 1967.)

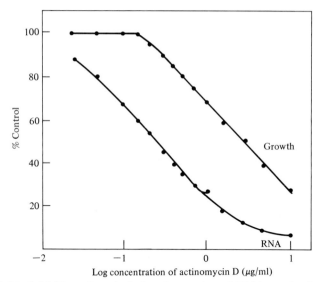

Figure 2-26. Inhibition of auxin-induced growth and RNA synthesis in soybean hypocotyl sections by actinomycin D. Low concentrations of actinomycin D inhibited RNA synthesis by as much as 30% without altering growth. (Redrawn, with permission, from Key et al., 1967.)

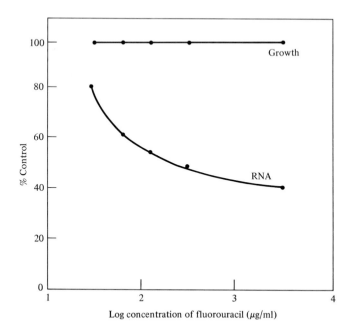

Figure 2-27. Inhibition of RNA synthesis but not auxin-induced growth by 5-fluorouracil. RNA synthesis was inhibited up to 60–70% without growth being affected. (Redrawn, with permission, from Key et al., 1967.)

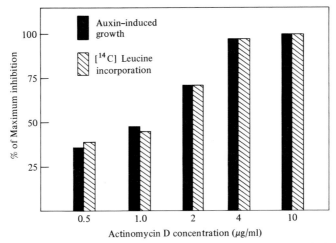

Figure 2-28. Inhibition of auxin-induced growth and protein synthesis in soybean hypocotyl sections by actinomycin D. (Redrawn, with permission, from Key et al., 1967.)

a more specific correlation was observed between inhibition of auxin-induced growth by actinomycin D and putative mRNA synthesis (Fig. 2-30). In those experiments, samples of elongating soybean hypocotyl sections were first preincubated for 4 hours in solutions containing 5-fluorouracil and various concentrations of actinomycin D. Then auxin and [8-¹⁴C]-ADP were added to the incubation solutions and auxin-induced growth and RNA synthesis were measured during the next 4 hours.

Since cell wall loosening is essential to auxin-induced growth, it was important to ascertain whether the auxin effect on wall loosening was in-

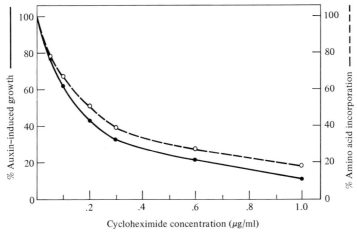

Figure 2-29. Parallel inhibition of auxin-induced growth and protein synthesis by cycloheximide. (Redrawn, with permission, from Key et al., 1967.)

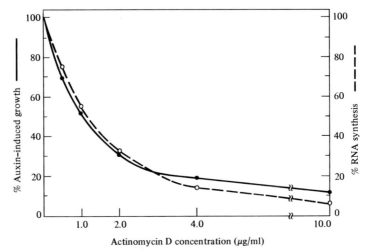

Figure 2-30. Parallel inhibition of auxin-induced growth and "D-RNA" (putative mRNA) synthesis by actinomycin D. 5-Fluorouracil also was present in the first incubation medium. (Redrawn, with permission, from Key et al., 1967.)

hibited to the same extent as inhibition of auxin-induced growth by inhibitors of RNA and protein synthesis. The results, displayed in Fig. 2-31, were positive.

The experimental data just discussed show that there is a general requirement for continued synthesis of RNA and protein, very probably mRNA and particular proteins, for continued cell enlargement in excised

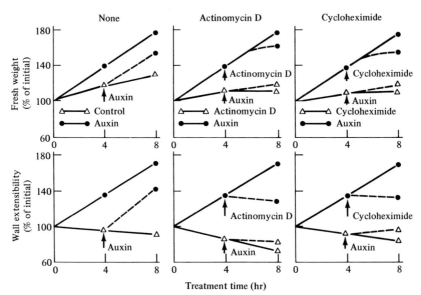

Figure 2-31. Effects of auxin, cycloheximide, and actinomycin D on growth and cell wall extensibility in soybean hypocotyl sections. (Redrawn, with permission, from Key et al., 1967.)

tissues over a period of several hours. The collective observations indicate that the ability of auxin to enhance the rate of cell enlargement is dependent upon new RNA and protein synthesis, and that auxin cannot cause a growth response by utilizing preexisting RNA or protein. The simplest interpretation of these observations, although not necessarily the correct one, is that the hormones are involved in the regulation of the synthesis of specific RNAs, probably mRNAs; that is, that the action of auxin is specific gene activation at the transcriptional level. These RNAs would then serve as templates for the synthesis of the proteins required for the physiological response in question. Let us examine additional evidence before drawing any such conclusions.

Refining the experimental system, investigators turned to investigations of the effects of auxin on chromatin-directed RNA synthesis using isolated nuclei and chromatin. It was observed very quickly in such investigations that chromatin isolated from 2,4-D-treated soybean hypocotyls had a greater than twofold higher RNA polymerase activity than control chromatin (Fig. 2-32). RNA polymerase of chromatin preparations showed no response to 2,4-D added to the chromatin directly. RNA polymerase from *Escherichia coli* greatly augmented RNA synthesis of chromatin preparations from both control and 2,4-D-treated plants, and both chromatin preparations exhibited similar saturation curves (Fig. 2-33). The saturation studies with *E. coli* RNA polymerase showed that only a small proportion of the available templates of chromatin from con-

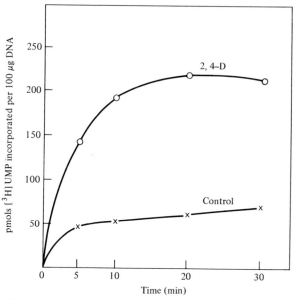

Figure 2-32. RNA synthesis by isolated chromatin and endogenous RNA polymerase from hypocotyls of control and 2,4-D-treated soybean seedlings. Seedlings were sprayed with a 2,4-D solution (1 mg/ml) 12 hours prior to chromatin isolation. (Redrawn, with permission, from O'Brien et al., 1968.)

Figure 2-33. Experiment as in Fig. 2-32 except the chromatin preparations from control and 2,4-D-treated seedlings were saturated with exogenous RNA polymerase. (Redrawn, with permission, from O'Brien et al., 1968.)

trol (7%) and 2,4-D-treated (16%) seedlings was transcribed in vitro by endogenous RNA polymerase. This result caused O'Brien et al. (1968) to suggest that the major influence of auxin was to increase the endogenous RNA polymerase of the isolated chromatin. However, the actual amount of RNA synthesis (picomoles) occurring in response to auxin was as great in the presence of added *E. coli* polymerase as with endogenous polymerase. This would seem to indicate that some difference exists between chromatin from control and auxin-treated tissue in addition to, or instead of, an effect on RNA polymerase. Additional evidence, obtained in 1970, showed that the RNA which was synthesized by chromatin from control and auxin-treated hypocotyl tissues were different, indicating that there is some difference in the portion of the genome being read. This could result from a change in the DNA template available for transcription or from association of RNA polymerase with template previously available but not being transcribed because of deficiency of RNA polymerase.

Related investigations were conducted by Matthysse and Phillips in 1969 on effects of auxin on RNA synthesis by isolated nuclei and isolated chromatin. They introduced for the first time, as regards plant hormones, evidence for a hormone receptor. Specifically, they found evidence for a nuclear protein factor which is required as an intermediate in the effect of auxin on RNA synthesis. If nuclei from tobacco and soybean were isolated in the presence of auxin, an auxin enhancement of RNA synthesis was obtained, upon treating nuclei with 3×10^{-7} M IAA. In the absence of auxin, a protein substance required for the hormone response apparently was lost from the nucleus. In a pea chromatin system fortified with *E. coli* polymerase, neither the factor nor auxin alone affected RNA synthesis. But when both auxin and the protein factor were added together, the chromatin showed a twofold increase in capacity for RNA synthesis. In the presence of this mediator or hormone receptor, auxin increased the rate of RNA synthesis by both isolated plant nuclei and by isolated chromatin; this increased rate of RNA synthesis occurred even in the presence of saturating amounts of (*E. coli*) polymerase. The hormone and protein did not affect the rate of RNA synthesis if pure DNA was used as the template. Matthysse and Phillips concluded that auxin interacts with a binding protein, the IAA–protein complex then interacting with chromatin to cause an increase in DNA template available for transcription. More will be said later about auxin receptors.

Rapid Responses to Auxin

Now let us turn to consideration of some responses of tissues to auxins which occur too rapidly and under circumstances which preclude a direct dependence on any primary gene activation effect of auxin. Examples of very rapid responses to auxin include: (1) elongation of coleoptile and stem sections, which can be observed to start in less than 15 minutes after presentation of a growth-promoting concentration of auxin; (2) protoplasmic streaming, which, in response to IAA, appears in 2 minutes or less; and (3) increase in respiration rate, which often occurs about 30 minutes after addition of auxin.

The so-called "rapid responses" to auxin generally are believed to be evident in less time than is required for transcription, translation, and action of a gene product (enzyme). Transcription alone requires on the order of 2 to 10 minutes, based on studies of enzyme induction and repression in bacteria. And translation, that is, polypeptide assembly on the ribosomes, takes an additional 1.5 to 5 minutes.

M. L. Evans and P. M. Ray published a paper in 1969 which provided impetus for a surge of interest in the rapid responses to auxin. They devised an experimental apparatus (Fig. 2-34) with which elongation of a file of several sections of stem or coleoptile could be measured accurately in short periods of time, and which permitted rapid changes of solutions bathing the sections. The results are recorded automatically on photographic paper attached to a rotating cylinder by a shadowgraph technique.

A typical result is illustrated in Fig. 2-35. Immediately after being mounted in the apparatus, the sections showed a burst of rapid elongation, due to tactile stimulation, which lasted for about half an hour. After the addition of auxin in growth-promoting concentration to the bathing solution, the sections then continued a slow rate of elongation for about 10

Side view   Top view

Figure 2-34. Growth measurement apparatus devised by Evans and Ray for use in experiments on rapid cell elongation responses. A, overflow reservoir, gas exit vent, and funnel for filling chamber; B, 7-mm-o.d. glass tubing; C, spectrophotometer cuvette (10-mm square); D, drain outlet, Tygon tubing; E, baffle, with slit (U); F, roll of photographic paper; G, glass plug; H, screw clamp; J, diaphragm; K, kymograph drum; L, zirconium arc lamp; M, weight; N, upper edge of shadow of M; O, row of coleoptile segments; P, P', Teflon plugs; S, No. 25 syringe needle; T, thread. (Redrawn, with permission, from Evans and Ray, 1969.)

minutes (at 23°C), the so-called "lag period," after which time a rather sudden increase in elongation occurred. Within about 3 additional minutes, a steady-state rate of elongation in response to auxin developed. Upon withdrawal of auxin (at $5 \times 10^{-5}\ M$ IAA), a decline in the rate of elongation began in approximately 10 minutes, and after 50 minutes the

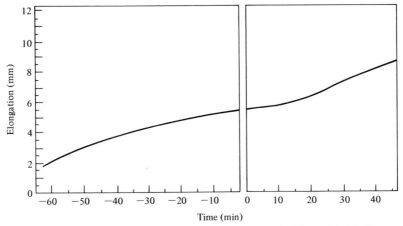

Figure 2-35. Record produced by the apparatus shown in Fig. 2-34. Medium was changed from water to a solution of IAA (3 μg/ml) at zero time. The elongation that was occurring at the beginning of the record was the result of tactile stimulation of the coleoptile sections as they were mounted in the apparatus. (Redrawn, with permission, from Evans and Ray, 1969.)

control rate was resumed. Later investigations have showed that the lag time in response to auxin could be reduced to less than 1 minute if the methyl ester of IAA, which penetrates cell membranes more rapidly than IAA, was used instead of IAA.

Very interestingly, neither actinomycin D nor cycloheximide at concentrations that partially inhibit the elongation response caused an extension of the lag period in the response to auxin, as would be expected if the rapid elongation was directly dependent upon de novo synthesis of RNA or protein.

Evans and Ray concluded in their 1969 paper that auxin probably does not act on the elongation of the tissues studied, oat and corn coleoptiles, by promoting the synthesis of informational RNA or enzymatic protein. They did not exclude the possibility that auxin acts at the translational level to induce synthesis of a structural protein, such as cell wall protein or membrane protein. However, they stressed that a number of alternatives to the gene-activation hypothesis of auxin action exist, any of which might depend indirectly on continual protein synthesis in order to remain functional. Serious consideration should be given to possible effects of auxin on intracellular transport processes or membrane permeability properties, for example, they suggested. And they emphasized another important point which should be borne in mind when using inhibitors of RNA and protein synthesis, namely, that antibiotics will inhibit the developmental process if it depends in any way upon RNA or protein synthesis, whether or not the hormone works by gene activation. It is only to be expected that enhanced growth will in general be accompanied by increased synthesis of RNA and protein, even if the hormone does not act directly upon these processes.

The "Acid-Growth" Theory of Auxin Action

Briefly stated, the "acid-growth" theory says that auxin initiates an acidification mechanism, possibly a membrane-bound H^+ pump, with the result that the pH of the solution in the matrix of the cell wall decreases. Some still hypothetical enzyme or enzymes reponsible for wall loosening are activated by the lowered pH, the wall is loosened, and cell enlargement takes place under the direct force of the cell's own turgor pressure.

This concept that hydrogen ions act as a second messenger for auxin-promoted cell elongation, formulated first by David L. Rayle and Robert Cleland in 1970, is supported by such evidence as: (1) segments of *Avena* coleoptile and pea and soybean stems elongate more rapidly at low pH; (2) wall extensibility in *Avena* and sunflower (*Helianthus*) is increased at low pH; (3) auxin-promoted elongation is preceded by auxin-induced H^+ extrusion in *Avena,* pea, and corn; and (4) certain wall-bound enzymes, glycosidases, have relatively low pH optima.

Auxin-induced H^+ excretion and the capacity of H^+ to promote growth of *Avena* coleoptile segments actually had been known for some time prior to formulation of the "acid-growth" theory. In 1934 James Bonner reported that the growth of coleoptile sections was eight times greater at pH 4.1 than at pH 7.2, and he also observed that a low pH induced a rather large increase in extensibility of the cell walls. And Kenneth V. Thimann noted in 1956 that elongation of coleoptile segments in response to auxin was accompanied by an acidification of the incubation medium. In the same year, 1956, Jean and Collette Nitsch reported that hydrogen ions had a stimulatory effect on cell enlargement both in the presence and absence of auxin.

If the wall-loosening factor is indeed H^+, certain criteria obviously must be met. One is that there ought to be similar growth responses to exogenous auxin and exogenous hydrogen ions. In the case of *Avena* coleoptile sections, this indeed is the case (Fig. 2-36). Sections begin to elongate rapidly about 10 minutes after the addition of auxin and reach a maximum rate in another 10–15 minutes. Likewise, the same maximum rate of elongation is obtained by an optimal H^+ concentration (pH 3.0), but the lag period is only about 1 minute. Both auxin and H^+ induce similar increases in wall extensibility, both agents are effective only when the cells' turgor pressure is nearly maximal, and both show the same unusual temperature dependence for elongation ($Q_{10} \simeq 5$ at 15–25°C and $Q_{10} \simeq 1$ between 25 and 35°C). Moreover, there are differences between the growth responses to auxin and H^+. Metabolic inhibitors (e.g., cyanide and dinitrophenol) prevent auxin-induced elongation but are without effect on

Figure 2-36. Kinetics of elongation of *Avena* coleoptile segments in response to 10^{-5} M IAA (pH 6.5 solution) and 10 mM citrate buffer at pH 3.0. (Redrawn, with permission, from Rayle, D. L. and R. Cleland. 1977. Control of plant cell enlargement by hydrogen ions. In: Moscona, A. A. and A. Monroy, eds. *Current Topics in Developmental Biology. Vol. 11. Pattern Development.* Pp. 187–214. Copyright by Academic Press, Inc., New York.)

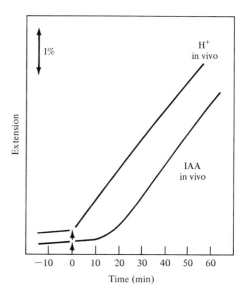

the acid-induced response, as the theory would predict. Furthermore, acidic solutions but not auxin cause wall loosening even when added to isolated cell walls.

Determination of the optimum concentration of H^+ for elongation presented some problems. The answer was found to depend really on whether or not the cuticle of the coleoptile segments was intact. With cuticle intact, sections extended only when the pH of the incubation solution was below 4.5, and the maximum response required a pH of 3. However, if the waxy cuticle (a barrier to H^+) was removed or abraded, cell wall loosening was induced whenever the pH was below about 5.8, and a maximum response occurred at pH 4.8 or below, a much more realistic physiological pH.

A second major criterion that must be satisfied to establish the validity of the acid-growth theory is that growth-promoting concentrations of auxin should cause excretion of H^+ from coleoptile cells. If coleoptile segments are "peeled" to remove the cuticle and floated on a small volume of a weakly buffered solution (e.g., 1 mM potassium phosphate, pH 6.0), and the pH is monitored with time, auxin-induced acidification can readily be detected within approximately 20 minutes (Fig. 2-37). If a flat surface electrode is placed directly on the *Avena* sections, acidification can be detected more rapidly. The acidification and growth responses induced by exogenous auxin (10^{-5} M IAA) obviously are directly correlated (Fig. 2-38).

More evidence for the role of H^+ as the wall-loosening factor came from studies with a fungal toxin called fusicoccin. Fusicoccin acts as a "super auxin," causing pea stem sections and *Avena* coleoptile sections

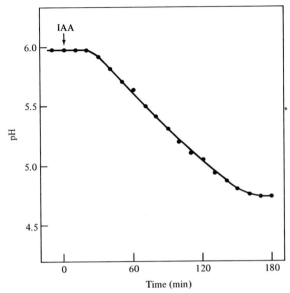

Figure 2-37. Auxin-induced acidification of a weakly buffered incubation medium (10^{-5} M IAA in 1 mM K-P$_i$, pH 6.0). (Redrawn, with permission, from Rayle, D. L. and R. Cleland. 1977. Control of plant cell enlargement by hydrogen ions. In: Moscona, A. A. and A. Monroy, eds. *Current Topics in Developmental Biology. Vol. 11. Pattern Development.* Pp. 187–214. Copyright by Academic Press, Inc., New York.)

to grow and excrete H$^+$ at rates greater even than those obtained with maximal auxin (Fig. 2-39). Thus two very different agents, IAA and fusicoccin, with different modes of action, both induce acidification with kinetics which closely approximate their relative speeds of action with respect to extension growth. The important conclusion, of course, is that

Figure 2-38. Effects of auxin (10^{-5} M IAA, added at zero time) on acidification of the incubation medium, **a**, and on elongation, **b**, of *Avena* coleoptile segments. (Redrawn with modifications, by permission, from Rayle, 1973.)

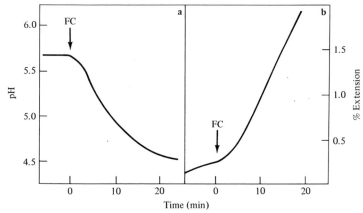

Figure 2-39. Kinetics of fusicoccin (FC)-initiated proton excretion, **a**, and growth, **b**, in *Avena* coleoptile sections. (Redrawn, with permission, from Rayle, D. L. and R. Cleland. 1977. Control of plant cell enlargement by hydrogen ions. In: Moscona, A. A. and A. Monroy, eds. *Current Topics in Developmental Biology. Vol. 11. Pattern Development.* Pp. 187–214. Copyright by Academic Press, Inc., New York.)

both cause H^+ extrusion, and it is the H^+ which evidently lead to wall loosening and thus to cell elongation.

The mechanism by which auxin causes wall acidification is unknown, but there are two hypothetical models worthy of consideration. According to one model, it is speculated that the primary action of auxin is at the plasmalemma. In 1972, J. W. Hardin, J. H. Cherry, D. J. Morré, and C. A. Lembi at Purdue University proposed that plasma membranes of soybean cells contain a regulatory factor which is specifically released by auxin, and which enhances the activity of purified soybean hypocotyl RNA polymerase. While by this model, the primary interaction of auxin would be with the plasmalemma, the Purdue group devised their hypothesis so as to account also for auxin-regulated nucleic acid biosynthesis (Fig. 2-40). Although still highly speculative, this model provides a possible mechanism whereby the interaction of auxin with the plasmalemma could be transmitted to the nucleus via the release of a receptor that could interact with a specific RNA polymerase. A result of this action would be altered DNA transcription and quantitative and qualitative changes in RNA synthesis. With a bit more speculation one could hypothesize that interaction of auxin with the receptor is accompanied by a release of H^+ from the plasmalemma into the cell wall, which effect seems to be directly related to cell wall loosening.

Ray (1977) reported on binding sites with high affinity for auxin on rough endoplasmic reticulum (ER) of maize coleoptiles, and he suggested an alternative model of auxin action in which the primary auxin action

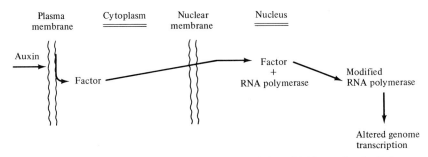

Figure 2-40. Hypothesis for auxin-regulated nucleic acid biosynthesis. It is proposed that auxin interacts with a factor within the plasma membrane. The factor is then transported through the nuclear membrane into the nucleus where it regulates the activity of RNA polymerase. (Redrawn, with permission, from Hardin et al., 1972.)

occurs at ER membranes. He reasoned that combination of auxin with its receptor sites on the ER could induce hydrogen ion transport from the cytoplasm into the cisternal space of the ER. The H^+ contained therein could be transported, along with secretory proteins contained in the ER space, to the cell wall, probably via the Golgi system. Assuredly, this model of auxin action via receptors on the ER suggests a possible connection between rapid and longer term effects of the hormone. As Ray (1977) stated, "Longer term auxin effects that involve stimulation of protein synthesis might stem from improved delivery of nascent secretory protein into the ER lumen, which transport might be coupled to the secretion of H^+ across the same membranes."

No discussion of the acid-growth theory would be complete, at the present state of our knowledge, without mention that there are some yet unresolved reports which are not in agreement. Two reports in particular, by L. N. Vanderhoef and associates in 1977, will be mentioned. Using excised elongating soybean (*Glycine max*) hypocotyl segments, Vanderhoef et al. (1977a,b) reported that elongating cells actively adjusted the external pH, from any pH in the range of 4 to 8, to pH 5.4. Auxin, at growth-promoting concentration, had no effect on the cellular pH adjustment of the external medium. From a comparison of acid-induced growth and auxin-induced growth, it was concluded that H^+ do not mediate in long-term auxin-induced elongation in soybean hypocotyl segments (Vanderhoef et al., 1977a,b). Only time and more research will enable possible reconciliation of these observations with the acid-growth theory. Meanwhile, the acid-growth theory will likely continue indefinitely to dominate research on the mechanism of auxin action. Which model most closely approximates what auxin really does? Only time and more investigation will enable us to say.

References

Ashton, F. M. and A. S. Crafts. 1973. *Mode of Action of Herbicides*. John Wiley & Sons, New York.

Audus, L. J. 1972. *Plant Growth Substances. Vol. 1. Chemistry and Physiology*. Leonard Hill Books, London.

Bandurski, R. S. and A. Schulze. 1974. Concentrations of indole-3-acetic acid and its esters in *Avena* and *Zea*. *Plant Physiol.* **54**: 257–262.

Beyer, E. M., Jr. 1972. Auxin transport: a new synthetic inhibitor. *Plant Physiol.* **50**: 322–327.

Bonner, J. 1934. The relation of hydrogen ions to the growth rate of the *Avena* coleoptile. *Protoplasma* **21**: 406–423.

Bonner, J. 1961. On the mechanics of auxin-induced growth. In: *Plant Growth Regulation*. Iowa State University Press, Ames. Pp. 307–328.

Bonner, J. and A. W. Galston. 1952. *Principles of Plant Physiology*. W. H. Freeman and Company, San Francisco.

Chang, Y.-P. and W. P. Jacobs. 1972. The contrast between active transport and diffusion of indole-3-acetic acid in *Coleus* petioles. *Plant Physiol.* **50**: 635–639.

Cleland, R. 1971. Cell wall extension. *Ann. Rev. Plant Physiol.* **22**: 197–222.

Cleland, R. E. 1976. Kinetics of hormone-induced H^+ excretion. *Plant Physiol.* **58**: 210–213.

Darwin, C. 1880. *The Power of Movements in Plants*. D. Appleton and Company, New York.

Davies, P. J. 1973. Current theories on the mode of action of auxin. *Bot. Rev.* **39**: 139–171.

dela Fuente, R. K. and A. C. Leopold. 1966. Kinetics of polar auxin transport. *Plant Physiol.* **41**: 1481–1484.

dela Fuente, R. K. and A. C. Leopold. 1972. Two components of auxin transport. *Plant Physiol.* **50**: 491–495.

Erdmann, N. and U. Schiewer. 1971. Tryptophan dependent indoleacetic-acid biosynthesis from indole, demonstrated by double-labelling experiments. *Planta* **97**: 135–141.

Evans, M. L. 1974. Rapid responses to plant hormones. *Ann. Rev. Plant Physiol.* **25**: 195–223.

Evans, M. L. and P. M. Ray. 1969. Timing of the auxin response in coleoptiles and its implications regarding auxin action. *J. Gen. Physiol.* **53**: 1–20.

Galston, A. W. 1967. Regulatory systems in higher plants. *Am. Scientist* **55**: 144–160.

Galston, A. W. and P. J. Davies. 1970. *Control Mechanisms in Plant Development*. Prentice-Hall, Englewood Cliffs, New Jersey.

Goldsmith, M. H. M. 1968. The transport of auxin. *Ann. Rev. Plant Physiol.* **19**: 347–360.

Goldsmith, M. H. M. 1977. The polar transport of auxin. *Ann. Rev. Plant Physiol.* **28**: 439–478.

Gordon, S. A. 1961. The biogenesis of auxin. In: Ruhland, W., ed. *Handbuch der Pflanzenphysiologie. Vol. XIV*. Springer-Verlag, Berlin. Pp. 620–646.

Gove, J. P. and M. C. Hoyle. 1975. The isozymic similarity of indoleacetic acid oxidase to peroxidase in birch and horseradish. *Plant Physiol.* **56**: 684–687.

Guilfoyle, T. J., C. Y. Lin, Y. M. Chen, R. T. Nagao, and J. L. Key. 1975. Enhancement of soybean RNA polymerase I by auxin. *Proc. Natl. Acad. Sci. U.S.A.* **72**: 69–72.

Haagen-Smit, A. J., W. B. Dandliker, S. H. Wittwer, and A. E. Murneek. 1946. Isolation of 3-indoleacetic acid from immature corn kernels. *Am. J. Bot.* **33**: 118–120.

Hager, A., H. Menzel, and A. Krauss. 1971. Versuche und Hypothese zur Primärwirkung des Auxins beim Streckungswachstum. *Planta* **100**: 47–75.

Hardin, J. W., J. H. Cherry, D. J. Morré, and C. A. Lembi. 1972. Enhancement of RNA polymerase activity by a factor released by auxin from plasma membrane. *Proc. Natl. Acad. Sci. U.S.A.* **69**: 3146–3150.

Hinman, R. L. and J. Lang. 1965. Peroxidase-catalyzed oxidation of indole-3-acetic acid. *Biochemistry* **4**: 144–158.

Keegstra, K., K. W. Talmadge, W. D. Bauer, and P. Albersheim. 1973. The structure of plant cell walls. III. A model of the walls of suspension-cultured sycamore cells based on the interconnections of the macromolecular components. *Plant Physiol.* **51**: 188–196.

Kende, H. and G. Gardner. 1976. Hormone binding in plants. *Ann. Rev. Plant Physiol.* **27**: 267–290.

Key, J. L. 1969. Hormones and nucleic acid metabolism. *Ann. Rev. Plant Physiol.* **20**: 449–474.

Key, J. L., N. M. Barnett, and C. Y. Lin. 1967. RNA and protein biosynthesis and the regulation of cell elongation by auxin. *Ann. New York Acad. Sci.* **144**: 49–62.

Kögl, F. and Haagen-Smit, A. J. 1931. Über die Chemie des Wuchsstoffs. K. Akad. Wetenschap. Amsterdam. *Proc. Sect. Sci.* **34**: 1411–1416.

Krul, W. R. 1972. Polar indole-3-acetic acid diffusion in nonliving and model systems. *Plant Physiol.* **50**: 784–787.

Labarca, C., P. B. Nicholls, and R. S. Bandurski. 1966. A partial characterization of indoleacetylinositols from *Zea mays*. *Biochem. Biophys. Res. Commun.* **20**: 641–646.

Lamport, D. T. A. 1970. Cell wall metabolism. *Ann. Rev. Plant Physiol.* **21**: 235–270.

Larsen, P. 1951. Formation, occurrence, and inactivation of growth substances. *Ann. Rev. Plant Physiol.* **2**: 169–198.

Leopold, A. C. 1955. *Auxins and Plant Growth*. University of California Press, Berkeley and Los Angeles.

Leopold, A. C. and O. F. Hall. 1966. Mathematical model of polar auxin transport. *Plant Physiol.* **41**: 1476–1480.

Leopold, A. C. and P. E. Kriedemann. 1975. *Plant Growth and Development*. 2nd ed. McGraw-Hill Book Company, New York.

Libbert, E., S. Wichner, U. Schiewer, H. Risch, and W. Kaiser. 1966. The influence of epiphytic bacteria on auxin metabolism. *Planta* **68**: 327–334.

Marré, E., P. Lado, F. Rasi-Caldogno, R. Colombo, M. Cocucci, and M. I. de Michelis. 1975. Regulation of proton extrusion by plant hormones and cell elongation. *Physiol. Vég.* **13**: 797–811.

Masuda, Y. 1969. Auxin-induced cell expansion in relation to cell wall extensibility. *Plant Cell Physiol.* **10**: 1–9.

Matheron, M. E. and T. C. Moore. 1973. Properties of an aminotransferase of pea (*Pisum sativum* L.). *Plant Physiol.* **52**: 63–67.

Matthysse, A. G. and C. Phillips. 1969. A protein intermediary in the interaction of a hormone with the genome. *Proc. Natl. Acad. Sci. U.S.A.* **63**: 897–903.

Miura, G. A. and S. E. Mills. 1971. The conversion of D-tryptophan to L-tryptophan in cell cultures of tobacco. *Plant Physiol.* **47**: 483–487.

Moore, T. C. 1969. Comparative net biosynthesis of indoleacetic acid from tryptophan in cell-free extracts of different parts of *Pisum sativum* plants. *Phytochemistry* **8**: 1109–1120.

Moore, T. C. and C. A. Shaner. 1967. Biosynthesis of indoleacetic acid from tryptophan-^{14}C in cell-free extracts of pea shoot tips. *Plant Physiol.* **42**: 1787–1796.

Moore, T. C. and C. A. Shaner. 1968. Synthesis of indoleacetic acid via indolepyruvic acid in cell-free extracts of pea seedlings. *Arch. Biochem. Biophys.* **127**: 613–621.

Morré, D. J. and J. H. Cherry. 1977. Auxin hormone-plasma membrane interactions. In: Pilet, P. E., ed. *Plant Growth Regulation*. Springer-Verlag, New York. Pp. 35–43.

Murray, A. K. and R. S. Bandurski. 1975. Correlative studies of cell wall enzymes and growth. *Plant Physiol.* **56**: 143–147.

Newman, I. A. 1963. Electric potentials and auxin translocation in *Avena*. *Aust. J. Biol. Sci.* **16**: 629–646.

Newman, I. A. 1970. Auxin transport in *Avena*. I. Indoleacetic acid-^{14}C distributions and speeds. *Plant Physiol.* **46**: 263–272.

Nitsch, J. P. and C. Nitsch. 1956. Studies on the growth of coleoptile and first internode sections. A new, sensitive, straight-growth test for auxins. *Plant Physiol.* **31**: 94–111.

O'Brien, T. J., B. C. Jarvis, J. H. Cherry, and J. B. Hanson, 1968. Enhancement by 2,4-dichlorophenoxyacetic acid of chromatin RNA polymerase in soybean hypocotyl tissue. *Biochim. Biophys. Acta* **169**: 35–43.

Paleg, L. G. 1965. Physiological effects of gibberellins. *Ann. Rev. Plant Physiol.* **16**: 291–322.

Percival, F. W. and R. S. Bandurski. 1976. Esters of indole-3-acetic acid from *Avena* seeds. *Plant Physiol.* **58**: 60–67.

Phillips, I. D. J. 1971. *Introduction to the Biochemistry and Physiology of Plant Growth Hormones*. McGraw-Hill Book Company, New York.

Phillips, I. D. J. 1975. Apical dominance. *Ann. Rev. Plant Physiol.* **26**: 341–367.

Price, C. A. 1970. *Molecular Approaches to Plant Physiology*. McGraw-Hill Book Company, New York.

Raven, J. A. 1975. Transport of indoleacetic acid in plant cells in relation to pH and electrical potential gradients, and its significance for polar IAA transport. *New Phytol.* **74**: 163–172.

Ray, P. M. 1958. Destruction of auxin. *Ann. Rev. Plant Physiol.* **9**: 81–118.

Ray, P. M. 1960. The destruction of indoleacetic acid. III. Relationships between peroxidase action and indoleacetic acid oxidation. *Arch. Biochem. Biophys.* **87**: 19–30.

Ray, P. M. 1974. The biochemistry of the action of indoleacetic acid on plant growth. In: Runeckles, V. C., E. Sondheimer, and D. C. Walton, eds. *The

Chemistry and Biochemistry of Plant Hormones. Vol. 7. *Recent Advances in Phytochemistry.* Academic Press, New York, Pp. 93–122.

Ray, P. M. 1977. Auxin-binding sites of maize coleoptiles are localized on membranes of the endoplasmic reticulum. *Plant Physiol.* **59**: 594–599.

Ray, P. M., P. B. Green, and R. Cleland. 1972. Role of turgor in plant cell growth. *Nature (London)* **239**: 163–164.

Rayle, D. L. 1973. Auxin-induced hydrogen-ion secretion in *Avena* coleoptiles and its implications. *Planta* **114**: 63–73.

Rayle, D. L. and R. Cleland. 1970. Enhancement of wall loosening and elongation by acid solutions. *Plant Physiol.* **46**: 250–253.

Rayle, D. L. and R. Cleland. 1972. The *in vitro* acid-growth response: relation to *in vivo* growth responses and auxin action. *Planta* **104**: 282–296.

Rayle, D. L. and R. Cleland. 1977. Control of plant cell enlargement by hydrogen ions. In: Moscona, A. A. and A. Monroy, eds. *Current Topics in Developmental Biology. Vol. 11. Pattern Development.* Academic Press, New York. Pp. 187–214.

Rayle, D. L., M. L. Evans, and R. Hertel. 1970. Action of auxin on cell elongation. *Proc. Natl. Acad. Sci. U.S.A.* **65**: 184–191.

Rayle, D. L. and W. K. Purves. 1967a. Isolation and identification of indole-3-ethanol (tryptophol) from cucumber seedlings. *Plant Physiol.* **42**: 520–524.

Rayle, D. L. and W. K. Purves. 1967b. Conversion of indole-3-ethanol to indole-3-acetic acid in cucumber seedling shoots. *Plant Physiol.* **42**: 1091–1093.

Salisbury, F. B. and C. W. Ross. 1978. *Plant Physiology.* 2nd ed. Wadsworth Publishing Company, Belmont, California.

Schneider, A. and F. Wightman. 1974. Metabolism of auxin in higher plants. *Ann. Rev. Plant Physiol.* **25**: 487–513.

Schrank, A. R. 1951. Electrical polarity and auxins. In: Skoog, F., ed. *Plant Growth Substances.* University of Wisconsin Press, Madison. Pp. 123–140.

Scott, T. K. 1972. Auxins and roots. *Ann. Rev. Plant Physiol.* **23**: 235–258.

Scott, T. K. and W. R. Briggs. 1960. Auxin relationships in the Alaska pea (*Pisum sativum*). *Am. J. Bot.* **47**: 492–499.

Scott, T. K. and W. R. Briggs. 1963. Recovery of native and applied auxin from the dark-grown 'Alaska' pea seedling. *Am. J. Bot.* **50**: 652–657.

Scott, T. K. and M. B. Wilkins. 1968. Auxin transport in roots. II. Polar flux of IAA in *Zea* roots. *Planta* **83**: 323–334.

Sherwin, J. E. 1970. A tryptophan decarboxylase from cucumber seedlings. *Plant Cell Physiol.* **11**: 865–872.

Talmadge, K. W., K. Keegstra, W. D. Bauer, and P. Albersheim. 1973. The structure of plant cell walls. I. The macromolecular components of the walls of suspension-cultured sycamore cells with a detailed analysis of the pectic polysaccharides. *Plant Physiol.* **51**: 158–173.

Thimann, K. V. 1934. Studies on the growth hormone of plants. VI. The distribution of the growth substance in plant tissues. *J. Gen. Physiol.* **18**: 23–34.

Thimann, K. V. 1937. On the nature of inhibitions caused by auxins. *Am. J. Bot.* **24**: 407–412.

Thimann, K. V. 1956. Studies on the growth and inhibition of isolated plant parts. V. The effects of cobalt and other metals. *Am. J. Bot.* **43**: 241–250.

Thimann, K. V. 1963. Plant growth substances; past, present and future. *Ann. Rev. Plant Physiol.* **14**: 1–18.

Thimann, K. V. 1969. The auxins. In: Wilkins, M. B., ed. *The Physiology of Plant Growth and Development.* McGraw-Hill Publishing Company Limited, London. Pp. 1–45.

Thimann, K. V. 1977. *Hormone Action in the Whole Life of Plants.* University of Massachusetts Press, Amherst.

Thimann, K. V. and F. Skoog. 1934. On the inhibition of bud development and other functions of growth substance in *Vicia fava. Proc. Royal Soc., Ser. B., Biol. Sci. London* **114**: 317–339.

Torrey, J. G. 1976. Root hormones and plant growth. *Ann. Rev. Plant Physiol.* **27**: 435–459.

Travis, R. L. and J. L. Key. 1976. Auxin-induced changes in the incorporation of ³H-amino acids into soybean ribosomal proteins. *Plant Physiol.* **57**: 936–938.

Tuli, V. and H. S. Moyed. 1967. Inhibitory oxidation products of indole-3-acetic acid: 3-hydroxymethyloxindole and 3-methyleneoxindole as plant metabolites. *Plant Physiol.* **42**: 425–430.

Ueda, M., A. Ehmann, and R. S. Bandurski. 1970. Gas-liquid chromatographic analysis of indole-3-acetic acid myoinositol esters in maize kernels. *Plant Physiol.* **46**: 715–719.

Vanderhoef, L. N., J. S. Findley, J. J. Burke, and W. E. Blizzard. 1977. Auxin has no effect on modification of external pH by soybean hypocotyl cells. *Plant Physiol.* **59**: 1000–1003.

Vanderhoef, L. N., T. S. Lu, and C. A. Williams. 1977. Comparison of auxin-induced and acid-induced elongation in soybean hypocotyl. *Plant Physiol.* **59**: 1004–1007.

Varner, J. E. and D. T. Ho. 1976. Hormones. In: Bonner, J. and J. E. Varner, eds. *Plant Biochemistry.* 3rd ed. Academic Press, New York. Pp. 713–770.

Verma, D. P. S., G. A. Maclachlan, H. Byrne, and D. Ewings. 1975. Regulation and *in vitro* translation of messenger ribonucleic acid for cellulase from auxin-treated pea epicotyls. *J. Biol. Chem.* **250**: 1019–1026.

Vickery, L. E. and W. K. Purves. 1972. Isolation of indole-3-ethanol oxidase from cucumber seedlings. *Plant Physiol.* **49**: 716–721.

Weaver, R. J. 1972. *Plant Growth Substances in Agriculture.* W. H. Freeman and Company, San Francisco.

Went, F. W. 1928. Wuchsstoff und Wachstum. *Rec. Trav. Bot. Néer.* **25**: 1–116.

Went, F. W. and K. V. Thimann. 1937. *Phytohormones.* Macmillan Company, New York.

Winter, A. 1966. A hypothetical route for the biogenesis of IAA. *Planta* **71**: 229–239.

Zenk, M. H. and H. Scherf. 1963. D-Tryptophan in höheren Pflanzen. *Biochim. Biophys. Acta* **71**: 737–738.

Chapter 3

Gibberellins

Brief History of Discovery

All research on the gibberellins (GAs) actually stems from the work of E. Kurosawa, a Japanese plant pathologist working in Formosa, who generally is credited with having discovered GA in 1926. However, the gibberellin story actually had its beginning in the last decade of the nineteenth century.

In 1898, Konishi, a semiliterate Japanese farmer, dictated an agricultural book in which was contained the earliest known description of what is now called the "bakanae" ("foolish seedling") disease of rice. The most characteristic symptom of the disease is the appearance of tall, spindly plants. Also, leaf sheaths are longer; leaves are longer, narrower, and thinner; root growth and tillering are reduced; plants are chlorotic; and severe cases cause death.

From 1898 to 1910, Hori provided the first descriptions of the disease with its causal fungus. In 1926, Kurosawa, after several failures, succeeded in producing the bakanae effect in rice and maize seedlings by treating them with cell-free culture medium in which *Gibberella fujikuroi* had been grown (Fig. 3-1). In 1931, Wollenweber corrected the taxonomy of the pathogen and established that bakanae is caused by *Fusarium moniliforme* Sheld., the asexual or imperfect stage of the ascomycete *Gibberella fujikuroi* (Saw.) Wr. Actually, *F. moniliforme* is known to attack a wide variety of hosts, among the most important being maize, cotton, rice, and sugar cane.

T. Yabuta, at the University of Tokyo, assigned the name "gibberellin" to the active factor in *G. fujikuroi* culture filtrates in 1935, and in 1938 Yabuta and Y. Sumiki announced the isolation of two crystalline, biologically active substances, which they named "gibberellins A and B."

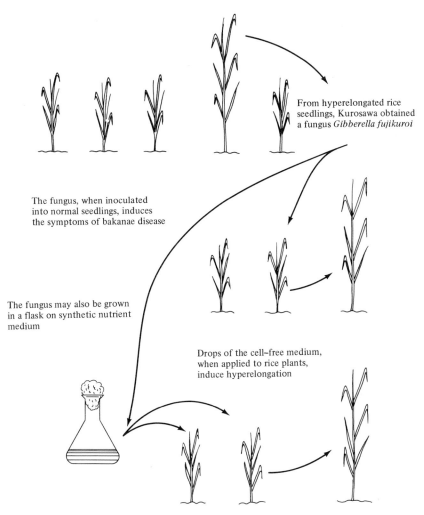

From hyperelongated rice
seedlings, Kurosawa obtained
a fungus *Gibberella fujikuroi*

The fungus, when inoculated
into normal seedlings, induces
the symptoms of bakanae disease

The fungus may also be grown
in a flask on synthetic nutrient
medium

Drops of the cell–free medium,
when applied to rice plants,
induce hyperelongation

The effective substance is gibberellin

Figure 3-1. Discovery of GA through investigations of bakanae, a disease of rice caused by *Gibberella fujikuroi* (*Fusarium moniliforme*). (Redrawn from *The Life of the Green Plant* by A. W. Galston, p. 74. Copyright © 1961 by Prentice-Hall, Inc. By permission of Prentice-Hall, Inc.)

It was nearly 10 years before Western abstracts of these papers became available. So actually, the existing GA story remained unknown in the Western world until about 1950, when several groups of workers, both in the United States and in England, discovered the existing Japanese papers.

In 1955, F. H. Stodola et al. of the U. S. Department of Agriculture undertook the purification of GAs from fungal filtrates. Apparently, almost simultaneously with the work of Stodola et al., scientists at Imperial

Chemical Industries in England (including P. W. Brian, A. Borrow, G. W. Elson, B. E. Cross, and others) began large scale preparation of GAs from fungus cultures. Both the American and British work resulted in the isolation of an entirely new compound from *G. fujikuroi* culture filtrates, named "gibberellin X" by the USDA team (1955) and "gibberellic" acid by the I.C.I. group (1954). The latter name has now been universally accepted, and gibberellic acid is now also known as GA_3.

Intensive research on the effects of gibberellic acid, of fungal origin, on higher plants was conducted in the 1950s, and in the same decade GA-like substances were discovered to occur naturally in higher plants.

The discovery that GAs are natural products of higher plants was made independently in 1956 by C. A. West and B. O. Phinney at the University of California at Los Angeles and by Margaret Radley of the Akers Research Laboratories of Imperial Chemical Industries in England (Radley, 1956; West and Phinney, 1956). GA-like substances were isolated from seeds and fruits of plants of nine genera representing seven families of angiosperms by Phinney and associates (West and Phinney, 1956; Phinney et al., 1957) and from pea (*Pisum sativum* L.) seedling shoots (Radley, 1956), and later from other parts of pea seedlings and mature seeds of several species, by Radley (1958). Actually, in 1951, J. W. Mitchell, D. P. Skaggs, and W. P. Anderson of the U. S. Department of Agriculture prepared extracts from immature bean seeds which had activity which can now be interpreted as GA activity, but at the time they did not know about GAs.

MacMillan and Suter apparently were the first to isolate and chemically identify a natural GA from higher plants when they extracted and identified GA_1 from immature seeds of *Phaseolus coccineus* L. in 1958. In the same year, West and Murashige isolated and identified GA_1 from seed of *Phaseolus vulgaris* L. Now it is firmly established that GAs occur naturally in a large number of species, and it is reasonable to conclude that these substances are ubiquitous in all angiosperms and gymnosperms, and plants of at least some other phyla in addition. The number of GAs now known from all sources is 52.

Chemical Characterization of GAs

GAs were defined by L. G. Paleg in 1965 as compounds having an *ent*-gibberellane skeleton (Fig. 3-2) and biological activity in stimulating cell division or cell elongation, or both, or such other biological activity as may be specifically associated with this type of naturally occurring substance.

GAs occur naturally in three chemical forms or states, two of which are chemically defined and the third of which is hypothetical: (a) "free GAs," (b) "conjugated GAs," and (c) other "water-soluble" or "bound GAs."

ent-gibberellane

Figure 3-2. Structural formula of *ent*-gibberellane.

Free GAs occur as the 19-carbon or 20-carbon, mono-, di-, or tricar-
boxylic acids unassociated by any detectable form of bonding to other
substances (Fig. 3-3). The terms C_{20}- and C_{19}-GAs denote compounds
which have retained and lost, respectively, carbon atom 20. Other differences among the GAs involve the presence or absence of a lactone configuration in ring A and the number and position of hydroxyl groups.

Many workers have compared the plant growth-promoting properties
of numerous GAs and some related compounds in a variety of bioassays.
It is now established that, besides differing in potency in any one test system, the GAs also exhibit a certain amount of plant species specificity.
P. W. Brian and associates at Imperial Chemical Industries in England examined 134 compounds related to GAs A_1–A_9 for growth-promoting activity in four bioassays in 1967, and they discussed extensively relationships between structure and activity. The general conclusions at that time
were: (a) no one particular feature of the active compounds could be regarded as essential for activity, but there appeared to be requirements for
the higher grades of activity; (b) the highly active compounds all
contained an intact gibberellane ring system (rings A, B, C, D), and it
seems safe to conclude that the intact gibberellane skeleton is essential for
growth-promoting activity; (c) the carboxyl group at C-7 is essential for
activity; and (d) the most active GAs tested all possess a lactone ring in
ring A.

There are numerous reports of the occurrence in fruits, seeds, potato
tubers, tulip bulbs, tomato seedlings, and tobacco shoot tips of GA-like
substances which are more polar than the GAs known prior to 1970.
These substances are more readily extracted with water or aqueous
buffers than with organic solvents such as methanol, which is commonly
used to extract free GAs. Unlike free GAs generally, these substances
cannot be efficiently extracted from acidified aqueous solution with ethyl
acetate but can be with *n*-butanol. In the early literature, such substances
were referred to as "water-soluble GAs" or, more commonly, "bound
GAs."

Several substances fitting the foregoing description have been isolated
and their structures determined in recent years. It is now recognized that

Figure 3-3. Structural formulas of 52 GAs. (Reproduced, with permission, from the *Annual Review of Plant Physiology,* Volume 29. © by Annual Reviews Inc. From Hedden et al., 1978.)

these comparatively polar forms of GA are chemically diverse, representing at least two, perhaps more, major types of substances. At least one recently described substance is, in fact, a highly polar "free GA." This is GA_{32}, which was isolated from immature seeds of *Prunus persica* L. and *Prunus armeniaca* L. and its structure determined. GA_{32} is reportedly active in eight biological systems. It was the most potent GA tested in the barley endosperm bioassay, being three times as active as GA_3.

Other relatively polar forms of GA which have been chemically characterized as natural constituents, specifically of certain seeds, in recent years are conjugates of β-D-glucose and particular GAs (Fig. 3-4). From immature seeds of Japanese morning glory (*Pharbitis nil* Chois.) six GA glucosides have been isolated and their structures determined. The

Figure 3-3 (continued).

GA₈ glucoside has also been isolated from immature seeds and seedlings of *Phaseolus coccineus* L. There have also been isolated at least four GA glucosyl esters (Fig. 3-4). Sembdner et al. termed the GA-glucosides "conjugated GAs" in 1968, and this name is likely to be widely adopted for the glucosides, glucosyl esters, and other conjugated forms of GAs that may be found. GA glucosides per se probably are inactive in growth regulation. Yokota and associates compared the growth-promoting effects of GA₃, GA₈, GA₂₆, GA₂₇, GA₂₉, and their glucosides in six bioassays in 1971. They found that the activities of the GA glucosides were much less than that of their aglycones, and concluded that the growth-promoting effects of the glucosides observed in some bioassay systems are probably due to the aglycones liberated by hydrolysis in plant tissue.

The term "bound GAs" continues to be used for as yet unidentified GA-like substances more polar than known GAs. Since free GAs report-

Figure 3-3 (continued).

edly are released in extracts containing unidentified "bound GAs" by treatment of extracts with preparations of proteinases, ficin and papain, for example, and by acid hydrolysis, some authors have suggested the existence of "protein-bound GAs." The reality of natural occurrence of GAs in protein-bound states remains uncertain, however. Whereas glucosyl-GAs have been isolated from tissues following treatment with free GAs, no such evidence for the formation of GA-protein complexes has been reported.

Currently, the major source of GAs for experimental and practical use is biosynthesis by large-scale cultures of various genetic strains of *Fusarium moniliforme*. However, progress has been and continues to be made to perfect organic synthesis of GAs. By 1969 the laboratory of Y. Sumiki in Japan had succeeded in the total synthesis of GA_2, GA_4, GA_9, and GA_{10} starting with o-xylene.

Gibberellin Glucosides

Figure 3-4. Structural formulas of GA glucosides and glucosyl esters. (Courtesy of N. Takahashi, 1976.)

Natural Occurrence of GAs

According to Anton Lang, by 1970, GAs and GA-like substances had been found in representatives of angiosperms, gymnosperms, ferns, brown algae, green algae, fungi, and bacteria. It would not be surprising, therefore, to ultimately discover that GAs are ubiquitous in the plant kingdom.

Selected examples of specific fruits and seeds from which particular GAs have been isolated are presented in Table 3-1. In general, it appears that a given organ or tissue contains two to several GAs. Furthermore, the kinds, amounts, and states (free or bound) of the GAs are not constant during development, as later discussion will show.

Attempts to determine the amounts of GAs in higher plant tissues generally have yielded at best only semiquantitative results. The bioassays employed often yield inexact data, and relatively few analyses have mea-

Table 3-1. Examples of Some Fruits and Seeds in Which
Specific GAs Have Been Identified[a]

Species	GAs
Marah macrocarpus (Greene) Greene	A_1, A_3
Mature seeds	A_4, A_7
Phaseolus coccineus L.	A_1, A_3, A_5, A_6
Immature seeds	A_8, A_{17}, A_{19}
Pharbitis nil Chois.	A_3, A_5, A_{20}
Immature seeds	A_{26}, A_{27}
Cucumis sativus L.	A_1, A_3, A_4
Immature seeds	A_7
Lupinus luteus L.	A_{18}, A_{19}
Immature fruits and seeds	A_{23}, A_{28}
Malus sylvestris L.	A_3, A_4, A_7
Fruits and seeds	

[a] With permission, from Moore and Ecklund (1975).

sured the amounts of identified GAs. Rather, the general practice has
been to estimate the amounts of unidentified GA-like substances present
in an extract by comparison with response of test material to standard
dosages of GA_3, or, less commonly, some other specific GA. Representative data are presented in Table 3-2. A general conclusion from the avail-

Table 3-2. Representative Data on Quantitative Analyses of GAs and GA-like
Substances in Plant Materials[a]

Plant material	Concentration of GAs	Assay methods
Immature florets of *Lolium perenne* L.	256 μg GA_3 equivalents per kg dry weight	Chromatography and lettuce hypocotyl bioassay
Immature seeds of *Pisum sativum* L.	1400 μg GA_3 equivalents per kg fresh weight; 0.430 μg per seed	Chromatography and dwarf pea and dwarf maize bioassays
Immature partheno-carpic fruit of *Malus sylvestris* Mill.	143 μg GA_3 equivalents per kg fresh weight	Chromatography, GLC, and dwarf pea and cucumber bioassay
Young fruits of *Citrus sinensis* Osbeck (9 days after petal fall)	Approximately 105 μg GA_3 equivalents per kg fresh weight; approx. 4 μg GA_3 equivalents per fruit	Chromatography and fluorometric analysis
7-day-old etiolated dwarf seedlings of *Pisum sativum* L.	88 μg GA_3 equivalents per kg dry weight; 0.033 μg per plant	Chromatography and dwarf maize bioassay

[a] With permission, from Moore and Ecklund (1975).

able data is that the concentrations of GAs in fruits and seeds, particularly immature seeds, are much higher, approximately two orders of magnitude in the latter, than in the vegetative organs of established herbaceous angiosperm sporophytes.

GA Biosynthesis in Seeds

Present knowledge concerning GA biosynthesis in higher plants is based largely on investigations of immature seeds. The biosynthetic pathway from mevalonate (Fig. 3-5), as far as it is currently known for higher plants, is based predominantly on the work by C. A. West and associates of the University of California at Los Angeles with preparations of liquid (free nuclear) endosperm of immature seeds of wild cucumber (*Marah macrocarpus*). Other systems which subsequently have been developed are cell-free enzyme systems from immature pea (*Pisum sativum* L.)

Figure 3-5. Pathway of GA biosynthesis from mevalonate as far as elucidated for seed plants.

seeds and fruits, immature pumpkin (*Cucurbita pepo* L.) seeds, seedling
shoots of pea and the castor bean (*Ricinus communis* L.), and sonicated
preparations of pea and *Brassica oleracea* L. chloroplasts.

Substantial progress toward the development of a cell-free enzyme system from higher plant tissues capable of GA biosynthesis first was made
when J. E. Graebe et al., working in C. A. West's laboratories, reported
in 1965 on the incorporation of [2-^{14}C]mevalonic acid into geranylgeraniol,
kaurene,[1] and kaurenol in extracts of endosperm-nucellus of immature
seeds of *Marah macrocarpus*. The intermediates in the formation of
geranylgeranyl pyrophosphate from mevalonate appear to be identical to
those involved in farnesyl pyrophosphate (hence sesquiterpene and triterpene) synthesis. Thus the isoprenoid pathway contributes to the synthesis
of abscisic acid (a sesquiterpenoid compound) and the isopentenyl side
chains of the cytokinins, as well as the GAs. In some cases the prenyl
transferase specificity determines the 20-C rather than the 15-C prenyl-
pyrophosphate as the major product of chain elongation. The cyclization
of geranylgeranyl pyrophosphate to kaurene occurs in two steps and
involves the formation of copalyl pyrophosphate as an intermediate. Both
reactions are catalyzed by kaurene synthetase.

Prior to 1975 the enzyme systems for GA biosynthesis were thought to
be cytoplasmic-microsomal although, as subsequent discussion will reveal, it is now clear that some if not all GA biosynthesis from mevalonate
is compartmentalized in plastids (see discussion of this topic also in Chapter 7). The reactions from mevalonate to kaurene are catalyzed by enzymes which are readily solubilized and which are dependent upon ATP
and a divalent cation in all systems which have been studied. In most systems the most effective cation is Mg^{2+}, but Mn^{2+} is more effective than
Mg^{2+} in cell-free extracts of immature pea seeds and pea seedling shoot
tips.

The further metabolism of kaurene consists of a series of oxidative
reactions which are catalyzed by mixed function oxidases and which are
dependent on oxygen and reduced pyridine nucleotide. The C-19 methyl
group is transformed to a carboxyl group as kaurenol, kaurenal and
kaurenoic acid are formed in sequence. Secondarily the resulting kaurenoic acid is hydroxylated in the 7β-position. T. C. Moore et al. presented
evidence in 1972 for the participation of a noncatalytic "carrier protein"
in facilitating reaction of kaurene with the oxidative enzymes in vitro in a
cell-free system from immature pea cotyledons.

---

[1] *Kaurene* refers specifically to *ent*-kaur-16-ene [also referred to as (−)-kaur-16-ene], and
kaurenol, kaurenal, and kaurenoic acid refer to *ent*-kaur-16-en-19-ol, *ent*-kaur-16-en-19-al,
and *ent*-kaur-16-en-19-oic acid, respectively. 7β-Hydroxykaurenoic acid refers to *ent*-kaur-
16-en-7αol-19-oic acid. The β-designation for the 7-OH group will be employed, in keeping
with previous publications (West, 1973). The trivial designation of the GAs and the *ent*-gib-
berellane system as described by MacMillan (1971) is employed.

The characteristics of the enzymes which catalyze the sequence of reactions leading from kaurene to 7β-hydroxykaurenoic acid have been studied primarily in the liquid endosperm of *Marah macrocarpus* seeds. All four reactions require reduced pyridine nucleotide and oxygen, which indicates that the enzymes are mixed function oxidases. The participation of cytochrome *P*-450 is indicated by the characteristics of carbon monoxide inhibition which shows maximal photoreversal at 450 nm and also interaction with other inhibitors. Regarding the requirement for reduced pyridine nucleotide, the best rates of oxidation have been consistently observed by supplying a mixture of NADH, NADP+, and ATP. Later, it was demonstrated by West and associates that the enzymes catalyze a pyridine nucleotide transhydrogenation reaction in which NADPH is formed at the expense of NADH in an ATP-dependent reaction.

More recently, evidence has been presented on the further conversion of 7β-hydroxykaurenoic acid to GAs in cell-free extracts of higher plant tissues. Contraction of the B ring of 7β-hydroxykaurenoic acid occurs by extrusion of the carbon atom in position 7. It is oxidative ring contraction involving the transfer of two electrons. Graebe et al. (1972) reported the formation of kaurenoic acid, 7β-hydroxykaurenoic acid, GA_{12}-aldehyde and GA_{12} from MVA in a cell-free system prepared from immature *Cucurbita pepo* L. seed endosperm. And Nakata *circa* 1972, in C. A. West's laboratories, obtained evidence for 7β-hydroxykaurenoic acid, $6\beta,7\beta$-dihydroxykaurenoic acid, GA_{12}-aldehyde, and GA_{12} as products of kaurenoic acid metabolism in lyophilized *Marah* endosperm preparations. While there is as yet no other evidence that GA_{12} occurs naturally in higher plants, M. Katsumi and B. O. Phinney proposed a scheme in 1969 whereby GA_{12} (or GA_{12}-aldehyde) could give rise to numerous GAs which are known to occur naturally in particular higher plants.

GA Biosynthesis in Systems Other Than Seeds

Prior to 1973, most investigations of GA biosynthesis in seed plants utilized cell-free enzyme extracts prepared from developing seeds and fruits. While these systems have yielded abundant valuable information about the biosynthetic pathway, the development of cell-free systems also from vegetative organs of young seedlings and older sporophytes is essential to an ultimate understanding of the factors regulating the kinds and amounts of GAs produced in plant tissues.

Synthesis of kaurene has been reported using cell-free extracts of pea (*Pisum sativum* L.) shoots by J. E. Graebe in 1968 and castor bean (*Ricinus communis* L.) seedlings by D. R. Robinson and C. A. West in 1970.

Table 3-3. Comparisons of Kaurene-Synthesizing Capacities of Cell-Free Extracts From Shoot Tips of Light- and Dark-Grown Alaska and Progress No. 9 Peas[a]

Alaska		Progress No. 9	
Light-grown[b]	Etiolated[c]	Light-grown[d]	Etiolated[e]
	dpm/mg protein		
78	17	40	35
	dpm/g fresh wt		
944	170	558	350

[a] With permission, from Moore and Ecklund (1974).
[b-e] Means of 5, 3, 4, and 1 independent experiments, respectively.

In 1973 the development of a cell-free system from pea shoot tips, the site of most abundant GA synthesis in the vegetative sporophyte, was reported which synthesized kaurene from MVA.

Kaurene is a key intermediate in the GA biosynthetic pathway, and it was considered plausible from previous results with developing pea seeds that the capacity of cell-free extracts of pea shoot tips to synthesize kaurene might be indicative of the potential of shoot tips to produce GAs. Accordingly, investigations were undertaken in an effort to ascertain possible correlations between rate of kaurene biosynthesis and rates of growth and other physiological parameters in etiolated and light-grown dwarf and tall cultivars of pea.

It was discovered in 1974 that cell-free extracts from shoot tips of light-grown Alaska pea (*Pisum sativum*) seedlings had a fivefold greater capacity for synthesizing *ent*-kaurene from mevalonate than extracts from shoot tips of etiolated seedlings of the same age (Table 3-3). Upon continuous irradiation of 10-day-old etiolated seedlings with high-intensity white light, activity increased approximately exponentially between the 3rd and 12th hours, attaining a level equal to that of light-grown plants of the same age, and remained nearly constant during the succeeding 24 hours of development (Fig. 3-6). Although the increase in capacity for *ent*-kaurene synthesis and chloroplast development occurred concurrently, there was no evidence then that the two processes were related.

These results reinforced the growing suspicion that at least some, if not all, of the observed synthesis of kaurene might be occurring in the chloroplasts. The author and associates had noted as early as 1974 that the procedures utilized to prepare enzyme extracts from pea shoot tips did not preclude contamination with enzymes or other substances of chloroplast origin, since the procedures resulted in extensive breakage of chloroplasts.

Hence, investigations were initiated to attempt to ascertain whether the enzyme kaurene synthetase occurs in pea chloroplasts. All the en-

Figure 3-6. Change in chlorophyll content and in kaurene-synthesizing capacity of Alaska pea shoot tips during 36 hour irradiation of 10-day-old seedlings with high-intensity white light, as measured in cell-free enzyme extracts. The drawings illustrate the shoot tips from which enzyme and chlorophyll extracts were prepared. (Redrawn, with permission, from Moore and Ecklund, 1974.)

zymes necessary for the synthesis of geranylgeranyl pyrophosphate from mevalonate were known to occur in chloroplasts in vivo, since the former compound is also a precursor of carotenoids. Thus the enzymic competence to cyclize geranylgeranyl pyrophosphate to kaurene was all that had to be shown in order to prove that chloroplasts possess complete metabolic competence for kaurene synthesis. It was gratifying to find and report in 1976 that enzyme extracts from sonicated pea chloroplasts could indeed enzymically convert geranylgeranyl pyrophosphate to kaurene. While that was in progress, P. D. Simcox, D. T. Dennis, and C. A. West reported in 1975 on the conversion of copalyl pyrophosphate to *ent*-kaurene in extracts of proplastids from etiolated pea shoot tips, developing castor bean endosperm extracts, and wild cucumber endosperm. Only the wild cucumber endosperm extracts were capable of converting GGPP to *ent*-kaurene.

The cumulative evidence to date seems clearly to warrant the conclusion that some if not all GA biosynthesis in higher plants, and perhaps some metabolic interconversions of GAs as well, is compartmentalized in chloroplasts and precursor organelles.

Interestingly, B. V. Milborrow reported in 1974 that abscisic acid, a sesquiterpenoid hormone (see Chapter 5), is synthesized from mevalonate by a preparation of lysed chloroplasts of ripening avocado fruit. Indeed in the avocado fruit, it appeared that the majority of the abscisic acid is synthesized within the chloroplasts, although the occurrence of synthesis outside chloroplasts was not discounted. T. W. Goodwin and colleagues have shown that there is a chloroplastic and an extrachloroplastic system for the conversion of mevalonate into terpenoid compounds. The chloroplast membrane is highly impermeable to mevalonate, and the mevalonate kinase of chloroplasts is distinct from the cytoplasmic mevalonate kinase. These observations reinforce the importance of plastids as compartments for hormone biosynthesis.

One naturally wonders why, during the course of evolution, synthesis of GAs and other terpenoid compounds would come to be compartmentalized in chloroplasts. All we can do at the present state of our knowledge is speculate. One possible explanation is that photosynthesis produces abundant adenosine triphosphate (ATP) and reduced pyridine nucleotide (NADPH), relatively large quantities of which are required for the biosynthesis of terpenoid compounds. Hence it would appear to be a positive adaptive feature for GA biosynthesis to be localized in an organelle where the levels of ATP and NADPH are comparatively high. There may also be a relationship between stimulation of GA biosynthesis by light and compartmentation in the chloroplasts.

Before leaving the subject of GA biosynthesis, mention should be made of some purely synthetic plant growth regulators (growth retardants). Several compounds have been produced which inhibit the growth of at least some seed plants, and which have been demonstrated to inhibit GA biosynthesis. Three compounds in particular have received attention both for their practical utility in inhibiting growth and for inhibiting GA biosynthesis. They are Amo-1618 [2'-isopropyl-4'-(trimethylammonium chloride)-5'-methylphenyl piperidine-1-carboxylate], CCC or Cycocel [(2-chloroethyl) trimethylammonium chloride], and ancymidol [α-cyclopropyl-α-(p-methoxyphenyl)-5-pyrimidine methyl alcohol] (Fig. 3-7).

Amo-1618 is the most specific inhibitor of GA biosynthesis in that it inhibits specifically the cyclization of geranylgeranyl pyrophosphate to copalyl pyrophosphate (see Fig. 3-5), one of the two reactions catalyzed by kaurene synthetase. However, Amo-1618 also inhibits the synthesis of sterols and other triterpenes. Specifically, it inhibits the cyclization step by which squalene-2,3-epoxide is converted to cyclized triterpene sterols.

CCC inhibits GA biosynthesis somewhere beyond the formation of kaurene. However, like Amo-1618 CCC also inhibits other reactions and

AMO–1618

[2'–Isopropyl–4'–(trimethylammonium chloride)–5'–methylphenyl piperidine–1–carboxylate]

$$CH_2Cl-CH_2-N^+\underset{CH_3}{\overset{CH_3}{-}}CH_3 \cdot Cl^-$$

CCC (Cycocel)

(2–Chloroethyl) trimethylammonium chloride

Ancymidol

[α–Cyclopropyl–α–(p–methoxyphenyl)–5–pyrimidine methyl alcohol]

Figure 3-7. Structural formulas of three synthetic inhibitors of GA biosynthesis.

in general appears to be the least specific of the two growth retardants. Ancymidol blocks the conversion of kaurene to kaurenol and the succeeding two mixed function oxidase-catalyzed reactions.

Effects of Light on GA Biosynthesis

There have been numerous reports of stimulation of GA biosynthesis by light, and in some cases phytochrome (see Chapter 7) has been implicated or directly shown to be involved as the photoreceptor. In the case of etiolated wheat leaf sections, an increase in GA-like substances occurred shortly after 5 minutes of red light treatment, which appeared to be the result of a release of GA from bound forms. However, in other work GA biosynthesis, as well as RNA and protein synthesis, appeared to be implicated in an observed increase in GA-like substances which occurred subsequent to a 30-minute red light treatment of wheat leaves. A transient increase in GA-like substances in etiolated barley leaf sections occurred within 15 minutes after a 30-minute exposure to red light and also appeared to require GA biosynthesis and RNA and protein synthesis. Additionally, D. M. Reid et al. reported in 1972 that homogenates of etiolated barley leaves, when exposed to brief periods (20 minutes or less) of red

irradiation, displayed an immediate increase in the levels of GA-like substances as compared to dark controls, and that the red light-induced increase in GA-like activity could be explained partially by the greater conversion of [³H]GA₉ into other GA-like substances with high biological
activity.

Previously an increase in kaurene-synthesizing activity in shoot tips of
Alaska pea seedlings during de-etiolation was described (Fig. 3-6). Since
kaurene is a key intermediate in GA biosynthesis, and rate of kaurene biosynthesis may be correlated closely with rate of GA biosynthesis, the
light-induced increase in kaurene-synthesizing capacity reflects a potential light-induced increase in rate of GA biosynthesis. In those investigations, continuous illumination for at least 12 hours was required to attain a
maximum rate of kaurene synthesis in greening shoot tips. This is a relatively sluggish light-induced change, compared to the results described in
the preceding paragraph. However, there also are reports of less rapid effects with which these observations agree more closely. D. Köhler reported in 1966 that exposure to red light [either continuous or intermittent
(2.5 minutes/hour)] resulted in an increase in extractable GA from shoots
of normal (tall) pea cultivar, and that the increase, susceptible to inhibition by CCC, appeared about 1 day after light exposure. Y. Ogawa
showed in 1964 that several hours of irradiation of etiolated *Pharbitis nil*
seedlings with white, red, or far-red light resulted in an increase in extractable GA-like substances, which was correlated with the opening of
the hypocotyl hook and expansion of cotyledons. On the basis of all available reports it seems certain that light may affect diverse reactions involved in GA metabolism. A more detailed discussion of these effects,
and the involvement of phytochrome in them, is presented in Chapter 7.

Whether there are significant differences in the levels of GAs in etiolated and light-grown seedlings of a given species and variety remains uncertain. R. L. Jones and A. Lang reported in 1968 that they detected no
significant difference between tall and dwarf peas or between etiolated
and light-grown peas with respect to their contents of extractable or
diffusible GA contents. On the contrary, D. Köhler found that the extractable GA content of light-grown tall or dwarf peas was considerably
greater than that of etiolated seedlings of the tall cultivar. Work by the
author and associates showed (Table 3-3) that enzyme extracts of the
shoot tips of light-grown seedlings of both a tall pea cultivar and a dwarf
pea cultivar possessed greater capacities for kaurene synthesis than enzyme extracts from their etiolated counterparts. These results are in
agreement with those of Köhler and, in certain respects also, with those
of G. W. M. Barendse and A. Lang who found in 1972 that a dwarf strain
of *Pharbitis nil* contained less GA than a tall strain, and that dark-grown
plants of both strains contained less GA than light-grown ones. Notably,
in this case the differences in GA content were attributable mainly to differences in levels of bound GA.

Interconversions of GAs

GAs differ markedly in their potency, and it appears that the particular GAs present in a tissue or organ at one time are not necessarily constant, that is, that metabolic interconversions occur. M. Katsumi and B. O. Phinney were first to present a scheme for interconversions among GAs, which was based on studies with *Gibberella fujikuroi*. More recently evidence has been obtained for higher plants, which agree generally with the data derived from studies with the fungus. Summarized in Fig. 3-8 are the reactions from kaurenoic acid to GAs and the known and postulated interconversions among the GAs. The interconversions based on experimental evidence are illustrated with solid lines, while the other postulated conversions based on comparative chemistry are indicated by dashed lines.

Among the salient features of Fig. 3-8 is that contraction of the B-ring of kaurenoic acid and succeeding 20-C intermediates results in an aldehyde which subsequently is oxidized to a carboxylic acid. Contraction of the B-ring is believed to take place prior to elimination of the carbon atom at position 7 and the formation of the C-4 → C-10 lactone characteristic of the 19-C GAs. Finally, there is abundant evidence that the 19-C GAs are formed from 20-C GAs.

Role of GAs in Dwarfism

For numerous species, particularly some of horticultural importance, there are genetic dwarf strains or cultivars and normal or tall types. In the case of the garden pea (*Pisum sativum* L.), for example, there are dozens of varieties or cultivars. Dwarf cultivars typically mature at heights of 30 cm or less (Fig. 3-9), whereas some tall or normal cultivars exceed a meter in shoot height at maturity.

Interestingly, dwarfism generally is readily apparent only in light-grown plants. When etiolated the difference in stem elongation between dwarf and tall cultivars is greatly reduced or nonexistent (see, e.g., Fig. 2-20). Furthermore, etiolated pea plants show little if any response to exogenous GA, whereas light-grown dwarf cultivars of peas and many other species often respond with dramatic stem elongation. Indeed it was in large part on the basis of this observation that it was first postulated that GAs are natural hormones of seed plants, and that GA deficiency might be causally related to genetic dwarfism.

In some cases, but not all, dwarfism does in fact seem to be correlated with GA deficiency. The case of dwarfism that is most completely understood at the molecular level is that of a dwarf mutant of corn (*Zea mays*) known as the dwarf-5 (d_5) mutant (Fig. 3-10). The d_5 phenotype is due to a single gene mutation, which results in such a marked reduction in inter-

Figure 3-8. Scheme for the biosynthetic pathways from kaurenoic acid to gibberellins and their interconversions. The conversions proven by feeding studies are shown by the heavy arrows and the structural formulas, and those based on comparative chemistry, are indicated by dashed arrows. (Courtesy of G. W. M. Barendse, 1976.)

node elongation that mature dwarf plants are only about 20% the height of normal plants. Dwarfism is expressed from the seedling stage to maturity and, in contrast to dwarf types of some other species, is expressed in etiolated as well as light-grown plants. The dwarf-5 mutant is deficient in endogenous GA, compared to normal genetic strains, and responds by nor-

Figure 3-9. Growth of dwarf peas (*Pisum sativum*) sprayed with GA 10 days after planting. The plant on the left was left untreated; the others were sprayed with increasing concentrations (10-fold increases), from 0.0015 to 15.0 mg/liter, progressing toward the right. Photographed 18 days after treatment. (Unpublished data of Moore, 1958.)

Figure 3-10. The effect of GA₃ on normal and dwarf corn (*Zea mays*). Left to right: normal control, normal plus GA, dwarf control, dwarf plus GA. (Courtesy of B. O. Phinney, 1978.)

(−)-Kaur–15–ene (isokaurene)   (−)-Kaur–16–ene (ent–kaurene)

Figure 3-11. Cyclization of copalyl pyrophosphate to the normal (−)-kaur-16-ene (ent-kaurene) and to (−)-kaur-15-ene (isokaurene), a deadend intermediate as in dwarf-5 mutant of corn (*Zea mays*). (Courtesy of Peter Hedden and B. O. Phinney, 1976.)

mal growth to appropriate dosages of exogenous GAs and such GA precursors as *ent*-kaurene, *ent*-kaurenol, and *ent*-kaurenoic acid.

In 1976, Peter Hedden and B. O. Phinney of the University of California at Los Angeles showed that the d_5 genetic lesion controls the cyclization of copalyl pyrophosphate (B activity of kaurene synthetase) to kaurene. The d_5 mutant produces mainly isokaurene (Fig. 3-11), an intermediate which is not converted to GAs, and only a comparatively small amount of the GA precursor *ent*-kaurene. Normal plants, on the other hand, efficiently synthesize *ent*-kaurene and only very small amounts of isokaurene (Fig. 3-12). Hence, dwarfism in the d_5 mutant is correlated, in a true cause and effect manner, with failure to carry out efficiently one reaction in GA biosynthesis.

Genetic dwarfs of some other species also appear to be deficient in GAs, compared to their normal or tall counterparts. Five dwarf cultivars of rice and one dwarf mutant of bean are reported to be GA-deficient. Yet in 10 other dwarf rice cultivars no differences in GA content, compared to normal plants, could be found.

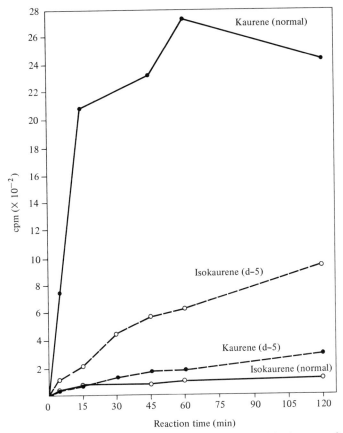

Figure 3-12. Comparative biosynthesis of *ent*-kaurene and isokaurene from [³H] copalyl pyrophosphate by cell-free enzyme extracts prepared from normal and dwarf-5 mutants of *Zea mays*. (Courtesy of Peter Hedden and B. O. Phinney, 1976.)

In 1972 G. W. M. Barendse and A. Lang reported on a comparison of the levels of endogenous free and bound GAs in the normal strain Violet and the dwarf strain Kidachi (Fig. 3-13) of Japanese morning glory. They found that the dwarf strain contains less GA than the tall counterpart, and that etiolated plants of both strains contained less GA than their light-grown counterparts (Fig. 3-14). In this case the differences in GA content reportedly were attributable mainly to differences in levels of bound GA.

The situation in peas is complex, according to the collective investigations reported on to date. H. Kende and A. Lang in 1964 found no difference in the extractable GA content in light-grown and dark-grown dwarf peas. Furthermore, R. L. Jones and A. Lang detected no significant difference between tall and dwarf peas or between light-grown and dark-grown peas with respect to their extractable or diffusible GA contents in a

Figure 3-13. Growth curves of the dwarf strain Kidachi and of the normal or tall strain Violet of Japanese morning glory (*Pharbitis nil*). Left, total height of shoot; right, fresh weight (of the whole plants). (Redrawn, with permission, from Barendse and Lang, 1972.)

study reported on in 1968. These authors described dwarfism in peas in terms of a light-induced difference between tall and dwarf cultivars in their sensitivity to a certain endogenous GA. However, this explanation is somewhat doubtful in view of other reports such as those of Köhler, who found that light-grown tall peas contained much more of a certain extract-able GA fraction than did light-grown dwarfs, and that the extractable GA content of light-grown tall or dwarf peas was considerably greater than

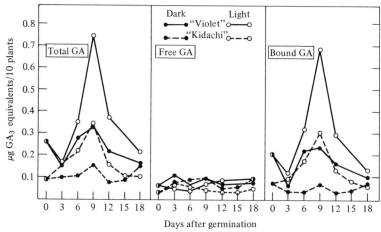

Figure 3-14. The GA content of light- and dark-grown plants of (dwarf) Kidachi and (normal or tall) Violet varieties of *Pharbitis nil* determined at 3-day intervals after germination. (Redrawn, with permission, from Barendse and Lang, 1972.)

that of the tall cultivar grown in the dark. P. R. Ecklund and T. C. Moore reported in 1974 that cell-free extracts from shoot tips of light-grown Alaska peas repeatedly possessed greater kaurene-synthesizing activity (1.4–5 times) than that found in extracts from light-grown Progress No. 9 shoot tips (Table 3-3). Furthermore, cell-free extracts from light-grown shoot tips of either the tall or the dwarf cultivar exhibited greater kaurene-synthesizing activity than was found in such extracts from etiolated shoot tips.

In view of the findings in recent years concerning free and bound GAs, and considering the extraction and bioassay techniques used in many studies in the past, many early investigations on levels of endogenous GAs in dwarf and normal plants need to be repeated, taking into account modern techniques and current knowledge. Until this is done, many conclusions about the role of GAs in dwarfism can be regarded only as tentative.

Other Aspects of GA Metabolism

Very little is known about the origin and the physiological significance of conjugated or bound GAs. With respect to physiological role, A. Lang (1970) stated that bound GAs can be considered either as reserve or storage forms, from which GAs are released, or as inert GAs which have no function, or as transport forms. The evidence for a role for conjugated GAs in GA transport rests on their occurrence in the bleeding sap of some trees. In bleeding sap from *Acer platanoides* and *Ulmus glabra*, treatment with a β-glucosidase caused a release of free GAs. Actually the two possible functions, serving as storage and as transport forms, are not mutually exclusive. In fact, the bleeding sap of maple and elm in which the conjugated GAs were found was collected in early spring, and the conjugates may represent storage GAs being transported to growth centers.

Only meager evidence is available concerning other metabolic conversions, including metabolic fate, of GAs in plants. Obviously, a comprehensive understanding of metabolic conversions of GAs is essential to an ultimate full understanding of the GA relationships of plants.

H. Kende prepared tritiated GA_1 and investigated its metabolism in dwarf peas in 1967. He applied $[^3H]GA_1$ to etiolated and light-grown dwarf peas and extracted the seedling shoots 24, 48, and 72 hours after treatment. There was some conversion of $[^3H]GA_1$ to an acidic, biologically active compound (approximately 40% of administered label after 72 hours), and a slight amount was converted to a neutral (water-soluble) fraction (about 10% after 72 hours).

There is abundant evidence for metabolic interconversions among free GAs and reversible formation of GA conjugates and other bound GAs during development of fruits, seeds, and other storage organs. Thus a free GA which is present at one time or at a particular developmental stage may not be present at some other time or stage of development.

It has been reported that developing seeds of *Phaseolus coccineus* L. (=*Ph. multiflorus* Lam.) contain GA_1, GA_3, GA_5, GA_6, GA_8, GA_{17}, GA_{19}, and GA_{20}, but so far none of these GAs have been demonstrated in mature seeds. Furthermore, the two major GAs which have been isolated from *P. coccineus* L. seedlings apparently are not identical with any of them. The 20-carbon GAs may in many cases be precursors of 19-carbon GAs, and interconversions among some 19-carbon GAs have been demonstrated.

Other evidence for metabolic interconversions or degradation of free GAs is scant. L. G. Paleg and B. G. Coombe found no evidence in 1967 for utilization of exogenous GA_3 in embryo-less barley half-grains during the induction of hydrolases in a barley endosperm bioassay procedure. A. J. McComb applied radioactive GA_3 to pea seedlings in 1964 and found no changes in the amount of recoverable GA_3 during a succeeding 94 hours. Some evidence has been reported for the metabolism of some exogenous GA_1 in seeds and seedlings, however, by H. Kende in 1967 and G. W. M. Barendse et al. in 1968.

The postulated storage function of bound GAs is indicated by evidence for interconversions between free and "bound" GAs during seed maturation and subsequent germination. G. W. M. Barendse et al., in a study published in 1968, injected radioactive GA_1 ([^{3}H]GA_1) into excised immature fruits of pea (*Pisum sativum* L.) and Japanese morning glory (*Pharbitis nil* Chois.) and investigated the fate of the applied GA_1 during further seed development and subsequent germination. During maturation of both kinds of seeds, part of the radioactivity became associated with the aqueous fraction, while another part of the [^{3}H]GA_1 was apparently converted to two other unidentified biologically active acidic compounds (Fig. 3-15). During germination of pea and morning glory seeds, part of the bound radioactivity was released in the form of [^{3}H]GA_1 and the two unidentified growth-active, acidic compounds. During rapid seedling growth, a further conversion of [^{3}H]GA_1, mainly to one of the unidentified acidic substances occurred.

It appears from these data that during seed development a part of the exogenous [^{3}H]GA_1 was converted to a bound form, while during germination part of this bound GA (approximately 10–15%) was reconverted to free GA. The latter process was accompanied by transfer of radioactive material from the cotyledons to the shoot–root axis. Whatever contribution of free GA may be accounted for by bound GA in early growth of pea seedlings, it appears that GA biosynthesis is initiated during early seedling growth and also that endogenous GA is present in growth rate-limiting quantity during early seedling development.

G. Sembdner et al. in 1968 applied radioactive GA_8 to developing *Phaseolus coccineus* L. pods and observed a decrease of radioactivity in

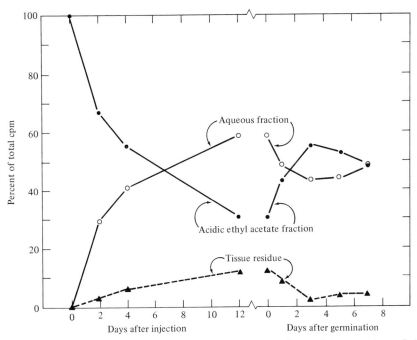

Figure 3-15. Distribution of radioactivity in extracts of pea (*Pisum sativum* L.) seeds developing in pods which were excised 13 days after anthesis, injected with [³H]GA₁, and grown in vitro, and in seedlings which subsequently developed from such seeds. The acidic ethyl acetate fraction contained free GAs and the aqueous fraction "bound GAs." (Redrawn, with permission, from Barendse et al., 1968.)

water-soluble substances and an increase in more ethyl acetate-soluble substances as the seeds germinated. These authors also observed transformations of labeled GA₃ and GA₆ following their application to maturing *Phaseolus coccineus* L. pods, the former to a hexapyranosyl-GA₃ and glucosyl-GA₈, the latter to GA₁/₃, GA₈, and glucosyl-GA₈. Barendse (1971) applied [³H]GA₁ to young *Pharbitis nil* Chois. plants and found that part of the exogenous GA was converted to two water-soluble compounds. The rate of conversion was enhanced when both [³H]GA₁ and [¹⁴C]glucose were applied to the same plants.

T. Hashimoto and L. Rappaport analyzed endogenous GAs in developing bean (*Phaseolus vulgaris* L.) seeds in 1966. They found that as the seeds matured the acidic ethylacetate-soluble GA-like substances initially decreased and then almost disappeared, while neutral substances and one of the acidic butanol-soluble substances increased. The neutral GA-like substances (bound GAs) were present in fairly large amounts in mature dry seeds. Related work by the same authors, involving the application of GA₁, further indicated formation of bound GAs during seed maturation, possibly products of conversion from acidic substances. In 1966 G. F. Pegg extracted GA-like substances possessing acidic, basic, and neutral properties from mature, dry tomato (*Lycopersicon esculentum* Mill.) seeds and young seedlings. He suggested that protein hydrolysis during germination is an important means of liberating free GAs from bound GAs.

Quantitative Changes in GA Content during Development

Development of Shoots

The author and P. R. Ecklund conducted studies in 1967 and 1968 designed in part to ascertain whether there are quantitative changes in sensitivity of shoots of pea plants of the normal or tall Alaska cultivar to exogenous GA and Amo-1618. It was reasoned that the relative rate of GA biosynthesis and the level of active, endogenous GA could be estimated indirectly by relating the growth rates to the growth responses of the plants to standard dosages of the two compounds.

In the initial phase of the investigations seedlings were treated with single growth-saturating doses of 10 μg of GA_3, which were applied to the shoot tips, at intervals of 3, 6, and 9 days after planting. Shoot height was measured daily from the third through the 13th day after planting. Seedlings treated at 3 days exhibited a pronounced stem elongation response (Fig. 3-16). Sensitivity of the seedlings to exogenous GA decreased as development progressed, to the extent that seedlings treated at 9 days of age were only slightly taller than the controls on the 13th day of growth. The plants completely failed to respond to exogenous GA when treated at 14 days of age. Hence, unlike dwarf varieties of peas, Alaska pea shoots become insensitive to exogenous GA at about 2 weeks of age.

Interestingly, however, Alaska pea plants again become sensitive to exogenous GA_3 after the linear phase of growth is passed and senescence begins. This was seen in experiments wherein plants were treated by applying 1 μg of GA_3 or 50 μg of Amo-1618. On some plants flowers and fruits were left on as they formed; the flowers were removed as they formed on other groups. Sensitivity to exogenous GA was measured as the percentage increase in stem elongation over that of untreated plants 1 week after application. Sensitivity to Amo-1618 was determined as the percentage inhibition in stem elongation compared with untreated plants 1 week after application. In Fig. 3-17 sensitivity to a growth regulator is plotted at the age of the plants when the regulator was applied. It was assumed that the growth response of a plant to exogenous GA would be inversely related to the level of active, endogenous GA in the shoot during the time the response is measured and that a specific amount of Amo-1618 would inhibit the biosynthesis of a specific amount of endogenous GA.

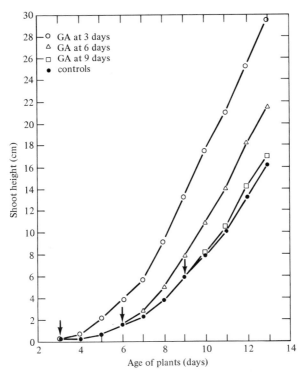

Figure 3-16. Time-course of stem elongation in pea seedlings to which single growth-saturating doses of GA, 10 μg per plant, were applied on the shoot tips on the days indicated by arrows. (Redrawn, with permission, from Moore, 1967.)

Therefore, the magnitude of growth inhibition by Amo-1618 would depend on the rate of GA biosynthesis throughout the time interval that inhibition is measured.

The plants displayed high sensitivity to both exogenous GA and Amo-1618 during the exponential phase of growth (Fig. 3-17), suggesting that endogenous GA was at a relatively low level, but a fairly high rate of GA biosynthesis was occurring. During the linear phase of growth, sensitivity to both exogenous GA and Amo-1618 declined, presumably because the rate of GA biosynthesis was rapidly increasing, thereby increasing the endogenous GA to a level that was practically growth-saturating.

In the case of plants with flowers and fruits intact, a second period of high sensitivity to exogenous GA appeared near the time of anthesis (Fig. 3-17) while sensitivity to Amo-1618 continued to decline, or remained quite low. These results suggest that subsequent to the linear growth phase the level of endogenous GA was decreasing because the rate of GA biosynthesis was rapidly declining. A lack of response to exogenous GA or Amo-1618 was associated with the rapid decline in growth rate, presumably because as apex senescence progressed, irreversible catabolic processes in the apex resulted in a loss of growth capability.

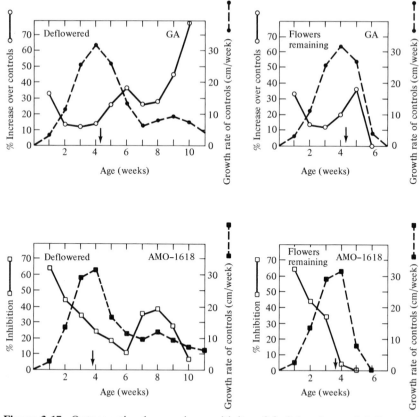

Figure 3-17. Ontogenetic changes in sensitivity of fruit-bearing and deflowered Alaska peas to a single shoot-tip application of 1 μg of GA₃ per plant or 50 μg of Amo-1618 per plant. Vertical arrows indicate the time of anthesis. Deflowered control plants were those from which flowers were removed as they formed and which received no growth regulator. Controls for experiments using plants with flowers remaining were flower- and fruit-bearing plants which received no growth regulator. Sensitivity to a growth regulator is plotted at the plant age at which the regulator was applied. Sensitivity to exogenous GA was measured as the percentage increase in stem elongation over that of untreated plants 1 week after the application. Sensitivity to Amo-1618 was measured as the percentage inhibition in stem elongation compared with untreated plants 1 week after the application. (Redrawn, with permission, from Ecklund and Moore, 1968.)

The decline in growth rate following anthesis was not as precipitous, the second period of high sensitivity to exogenous GA was delayed, and inhibition by Amo-1618 did not decrease as rapidly with deflowered plants as with flower- and fruit-bearing plants. These differences are assumed to result from a slower decline in the rate of GA biosynthesis following anthesis in deflowered plants. Throughout the period of prolonged growth due to flower removal, the sensitivity to exogenous GA was inversely related to the growth rate of the plants, while inhibition by Amo-1618 was directly related to the growth rate. The troughs in the GA-sensitivity curves correspond with the peaks in the Amo-1618-sensitivity curves for deflowered plants, suggesting that a second period of increasing GA biosynthesis occurred in the shoot tips of plants in which apex senescence was delayed. As deflowered plants approached the stage of determinate growth with the apparent disappearance of the shoot apex, the growth potential of the shoots was lost and the sensitivity to exogenous GA or Amo-1618 disappeared.

It appears from these experiments that light-grown Alaska peas exhibit two periods of sensitivity to exogenous GA in their ontogeny, the first being the growth period prior to the linear phase of growth, and the second being the period of declining growth rate and onset of apex senescence. Thus it seems that endogenous GA is rate-limiting for stem elongation in light-grown Alaska peas during early seedling growth and during the period of decline. It may be speculated that the level of endogenous GA in Alaska peas is regulated primarily by the rates of synthesis and utilization in growth processes. Stem elongation in light-grown Alaska peas may be specifically limited by the rate of endogenous GA biosynthesis prior to and subsequent to the linear phase of growth. Insensitivity to exogenous GA during the linear phase of growth probably is correlated with the presence of growth-saturating levels of endogenous GA. A direct approach toward confirmation of this hypothesis would be to conduct time-course studies of the rate of GA biosynthesis and the concentration of extractable GA in pea shoots throughout ontogeny.

To date precisely those types of investigations have not been reported. However, the author and P. R. Ecklund did conduct studies on ontogenetic variations in capacity for kaurene synthesis in enzyme extracts of Alaska pea plants. The results of that effort are illustrated in Fig. 3-18. Kaurene synthesis was not detectable in epicotyls of 3-day-old seedlings. However, activity was detectable in extracts of plants older than 3 days and became maximal in shoot tips of 9-day-old plants. Thereafter kaurene-synthesizing activity remained relatively constant until the 24th day, during which interval the growth rate was increasing. The same relationship can be seen when activity is expressed on a fresh weight basis. These results reveal a direct correlation between the apparent rate of kaurene synthesis in pea shoot tips and the apparent rate of GA production, as assessed by indirect procedures in previous investigations. Interestingly, the results are also analogous to those which were reported previously by R. C. Coolbaugh and the author for developing pea seeds, in which a large increase in capacity for kaurene production appears to slightly precede the period of maximum GA biosynthesis and maturation (see Fig. 3-20).

Figure 3-18. Ontogenetic changes in growth rate and in kaurene-synthesizing capacity of Alaska pea shoot tips, as measured in cell-free enzyme extracts. The drawings illustrate the shoot tips from which enzyme extracts were prepared. The data represent means based on three independent experiments. (Redrawn, with permission, from Moore and Ecklund, 1974.)

Development of Fruits and Seeds

As in the case of shoots, only fragmentary and incomplete evidence exists regarding changes in GA content, rate of GA biosynthesis, and correlations between GA content and growth rate in fruits, seeds, and other specific organs. However, developing seeds and fruits have been most intensively investigated, primarily because they contain much higher concentrations of GA than vegetative organs. Questions of interest include where in the seeds and fruits GA synthesis is localized, and how the levels of endogenous GA vary during development.

In immature seeds of many species of angiosperms the maximum amounts of extractable GAs are found when the seeds have attained about half their maximum fresh weights (e.g., Fig. 3-19). The characteristic decrease in extractable GA content as the seed matures quite probably is correlated with the formation of conjugated GAs or other forms of bound

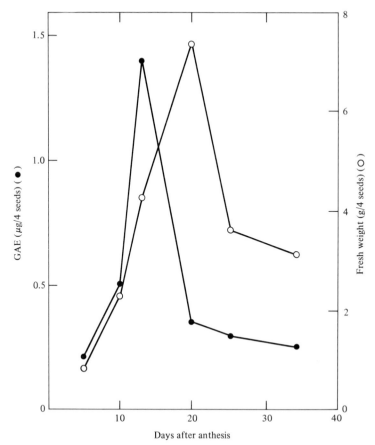

Figure 3-19. Changes in GA content and fresh weights in seeds of *Pharbitis nil* Chois. (Japanese morning glory) from anthesis to maturity. GA content is expressed as GA_3 equivalents (GAE). (Redrawn, with permission, from Murakami, 1961.)

GA which are not effectively extracted from acidified aqueous solution with relatively nonpolar solvents. By extraction procedures which used to be commonly employed, such substances, when present, generally were inadvertently discarded in an aqueous fraction.

It is interesting to compare these data with available information from studies with cell-free enzyme extracts. In 1969 R. C. Coolbaugh and T. C. Moore found that the capacity to convert mevalonic acid to kaurene in cell-free enzyme extracts varied markedly with the stage of seed development (Fig. 3-20). Kaurene-synthesizing capacity increased from a barely assayable level in very young ovules to a maximum at about 13 days after anthesis, when the seeds had attained about half-maximal fresh weight.

J. E. Graebe in 1968 observed comparatively low incorporation of MVA into kaurene in cell-free extracts prepared from very young pea

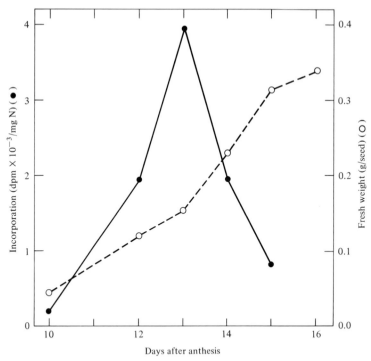

Figure 3-20. Net rates of [^{14}C]kaurene biosynthesis from [2-^{14}C]mevalonate in cell-free enzyme extracts (40,000 *g* supernatant fraction of homogenates) prepared from pea (*Pisum sativum* L.) seeds at different stages of development, in relation to seed growth. (Redrawn, with permission, from Coolbaugh and Moore, 1969.)

fruits and even lower incorporation with immature seeds from almost fully grown pods. The kaurene-synthesizing activity that he observed with very young fruits probably reflected synthesis in the carpel. Hence, just as the growth curves for the legume pod and seed do not coincide in time (Fig. 3-21), maturation of the seed lagging behind that of the pod, it seems likely that the time-courses of kaurene and GA synthesis and GA accumulation in the pod and seed are similarly related.

J. L. Stoddart in 1965 investigated changes in GA content during ripening of seeds of two species of grasses. Floret samples of perennial ryegrass (*Lolium perenne* L.) and timothy (*Phleum pratense* L.) were analyzed for GA content at 2–5 day intervals from inflorescence emergence to seed harvest, using chromatographic and bioassay techniques. Both species had a relatively high GA content at inflorescence emergence, which diminished during anther formation. During final seed maturation the level of GA decreased steadily to reach a stable, low level prior to harvest.

Changes in endogenous GAs during development of fleshy fruits have also been investigated rather extensively. D. I. Jackson and B. G.

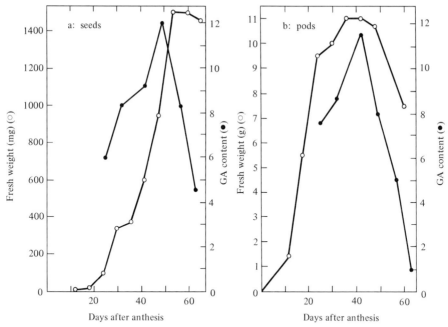

Figure 3-21. Changes in amount of GA-like substances and fresh weight in seeds, **a**, and pods, **b**, of bean (*Phaseolus vulgaris* L.) during development. GA content is expressed as mm of growth in the dwarf (d_1) *Zea mays* leaf section bioassay. (Adapted from Wareing and Seth, 1967, by permission.)

Coombe in 1966 compared growth rates of the seed, endocarp, and meso-carp of apricot (*Prunus armeniaca* L.) between anthesis and fruit maturity with the concentrations of GA-like substances extracted from those tissues, and also described some properties of the GA-like substances. A strong correlation was shown between the concentration of GA-like substances in all three tissues and their respective growth rates for the first 60 days after anthesis (Fig. 3-22). Results with methanol extracts showed that in all three parts of the apricot fruit, GA activity increased after anthesis and reached a maximum 20 days after anthesis. The seed had the greatest GA activity and the endocarp the least. Thereafter concentrations decreased in all three tissues, the decrease being most rapid in the endocarp and slowest in the mesocarp. The time at which GA activity was greatest coincided with the time of greatest growth rate in the mesocarp, endocarp, and nucellus. Activity in the endocarp declined to a low level concurrently with cessation of growth and sclerification of the tissue, 30–45 days after anthesis. Activity in the seed remained high and did not decline until 35–60 days after anthesis during which seed growth stopped. The only tissue that grew after 60 days was the mesocarp, which contained the highest content of GA-like substances at that time. Evidence for the occurrence of bound GAs, that is substances more polar

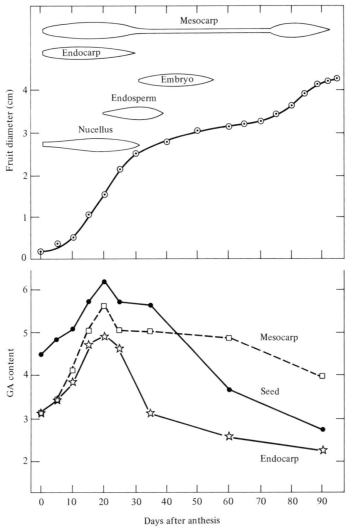

Figure 3-22. Concentrations of GA-like substances in methanolic extracts of seed, endocarp, and mesocarp of apricot (*Prunus armeniaca* L.) fruits in relation to their growth rates during development. GA content is expressed as log pg GA₃ equivalents/10 mg tissue determined by the barley endosperm bioassay. Figures with tissue names inserted in top half of figure represent the comparative volume increases of the tissues. (Redrawn, with permission, from Jackson, D. I. and B. G. Coombe. 1966. Gibberellin-like substances in the developing apricot fruit. *Science* **154**: 277–278, 14 October 1966. Copyright 1966 by the American Association for the Advancement of Science.)

than free GAs, in developing apricot fruits was seen in the fact that methanol extracts of the seed and pericarp were 10 times more active than ethyl acetate extracts.

D. I. Jackson reported in 1968 on an analogous, equally impressive study of changes in GA-like substances during development of peach (*Prunus persica* Batsch) fruits. No GA was found in the ovary before bloom. Immediately after full bloom GA activity was found first in the seed and later in the mesocarp and endocarp as well. GA concentration was closely correlated with the rate of cell expansion in each tissue, but not with cell division. Until the final growth phase (in the double-sigmoid growth curve), when activity was found only in the mesocarp, the highest GA level was always found in the seed. W. J. Wiltbank and A. H. Krezdorn reported in 1969 that there was a striking direct correlation between concentration of GA-like substances and mean fresh weight of whole navel orange (*Citrus sinensis* Osbeck) fruits during development (Fig. 3-23).

An interesting question concerns the role of GAs produced in seeds in the development of fleshy fruits. D. I. Jackson stated in 1968 that hormone levels in seeds rarely correlate with the growth of the whole fruit. More commonly, he noted, high levels of hormone correspond to periods of rapid seed growth. This is sometimes interpreted to indicate that growth of other fruit tissue is not directly dependent on supplies of hormone synthesized by the seed. D. I. Jackson and B. G. Coombe found in 1966, in fact, that in apricot the GA concentration may at times be higher in the mesocarp than in the seed and that the respective tissues of apricots may be self-sufficient for this hormone. Similar results with peach fruits supported that view and further suggested that GA levels in a given tissue are intimately associated with cell expansion in that tissue.

S. Iwahori et al. in 1968 determined levels of GA-like activity in berries of both seeded Tokay and Seedless Tokay grapes (*Vitis vinifera* L.) at different stages of development. In general, they found, as did R. J. Weaver and R. M. Pool in 1965, more GA-like activity in seeded than in seedless berries, indicating that grape seeds may be a rich source of GA-like substances. This view is supported by the fact that exogenous GA greatly promotes enlargement of Seedless Tokay berries, but not seeded Tokay berries. In Seedless Tokay the development of seeds ceases at an early stage. The authors stated that the earlier decrease of GA in the Seedless Tokay berries probably is associated with the abortion of the embryo and cessation of seed development.

Are GAs essential to the development of seeds? Ironically, an unequivocal affirmative answer may not yet be fully justified, although on the basis of available evidence it seems reasonable to conclude that GAs are indeed essential for the development of seeds, just as for the sporophytes which potentially develop from them. Seeds apparently accumulate GAs

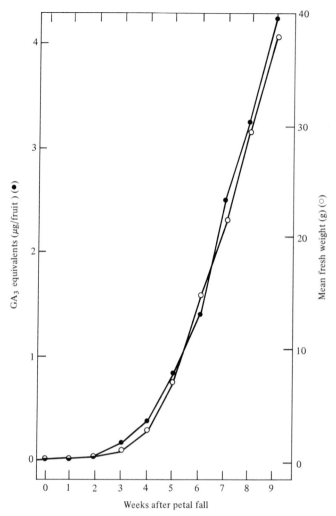

Figure 3-23. Correlation between concentration of GA-like substances and growth of navel orange (*Citrus sinensis* Osbeck) fruits in a 9-week period following petal fall. Concentrations of GA-like substances are expressed in terms of GA$_3$ equivalents determined by a fluorometric assay. (Redrawn, with permission, from Wiltbank and Krezdorn, 1969.)

in forms from which some free, active GAs potentially can be mobilized to the seedling shoot–root axis during germination. However, it is certain that far more GAs accumulate in seeds than is required for their own maturation.

In summary, whether GAs produced in seeds contribute to the growth of fleshy fruits in which they are contained is not clear from presently available evidence. Some data suggest that each tissue of a fleshy fruit has a relatively independent GA metabolism, while other evidence indicates that seeds in fleshy fruits do contribute GAs or GA precursors to the surrounding tissues. In general, very good, direct correlations have been found between GA content and growth of whole fleshy fruits and the constituent tissues thereof. A decrease in endogenous GAs appears to be associated with the ripening of fleshy fruits.

Sites of GA Biosynthesis in Seed Plants

J. A. Lockhart in 1957 provided the first evidence on the sites of GA biosynthesis using etiolated pea seedlings. His method of establishing the organ of production was to remove individual parts of the plant suspected of being the source and then attempt to reproduce the response normally elicited by the organ by application of pure hormone. His rationale was that, if the added substance can completely or to a large extent replace the organ in restoring normal growth, then the presumption is strong that the organ removed normally supplies a comparable substance to the plant. It may be noted that the converse conclusion is not valid. He showed that applied GA_3 could almost completely replace the stem tip in promoting the elongation of the subapical region of the pea stem and concluded that the natural GA of the pea plant is produced in the stem tip.

R. L. Jones and I. D. J. Phillips studied the sites of GA synthesis in light-grown sunflower plants in 1966 and 1967 using techniques of organ excision and application of GA, extraction, diffusion, and treatment with inhibitors of GA biosynthesis. They found that GA synthesis occurred in the young leaves (not the apical meristem per se) of the apical bud. Older but not fully grown leaves supply GA to the stem although in lower quantity. Root tips also were shown to be sites of GA synthesis, this ability being confined to the apical 3 to 4 mm of the root. The findings that the amounts of GAs obtained over a period of time exceeded the amounts of GAs obtained by extraction, and that CCC reduced the amounts of GAs obtainable over time by the diffusion technique, provided strong corroborative evidence for GA biosynthesis in the excised organs.

D. Sitton, A. Richmond, and Y. Vaadia provided additional evidence in 1967 for potential GA biosynthesis in sunflower root tips. They extracted GA-like substances from the roots of decapitated plants, and reported further that incubation of the root tips in [2-^{14}C]mevalonate yielded kaurenol, an intermediate in GA biosynthesis.

As indicated in the discussion of the biosynthetic pathway, developing seeds are rich sources of GAs and are comparatively intensely active in GA biosynthesis. Not only do developing seeds contain relatively large amounts of GAs, inhibitors of GA biosynthesis also can be demonstrated to inhibit accumulation of GA in seeds in situ.

Direct evidence on sites of localization of enzymes involved in GA biosynthesis in developing seeds is fragmentary and incomplete and is derived mainly from studies of immature seeds of two species, pea (*Pisum sativum* L.) and wild cucumber (*Marah macrocarpus*). R. C. Coolbaugh and T. C. Moore in 1971 performed comparative assays of kaurene accumulation in cell-free enzyme extracts of whole pea (cv. Alaska) seeds, isolated seed coats, cotyledons, and embryonic shoot–root axes. Kaurene, and presumably GA, biosynthesis from mevalonic acid was localized exclusively in the cotyledons.

These results differ rather strikingly from comparable data for wild cucumber seeds in which kaurene synthesis is localized in the liquid endosperm. This difference becomes intriguing when it is realized that the early morphological development of pea seeds also includes the formation of liquid endosperm. During early embryogenesis, the pea embryo occupies a small part of the embryo sac, the remainder being filled with liquid (free-nuclear) endosperm. By the time the seed is 11 to 12 days old, the endosperm is entirely digested, and the seed consists of the embryo and the seed coat. The pea seeds utilized in the investigations of R. C. Coolbaugh and T. C. Moore were 12 to 16 days of age.

Interesting changes in kaurene-synthesizing capacity occur during the development of both pea and wild cucumber seeds. In extracts of *Marah macrocarpus* the highest incorporation of mevalonic acid into a lipid fraction (containing kaurene) occurs in seeds when the cotyledons are not yet visible and the seed cavity is mostly filled with endosperm, and the activity declines as the seeds mature and the cotyledons form. Cell-free extracts of pea seeds exhibit a relatively low capacity for kaurene biosynthesis during early morphological development when liquid endosperm is present. In fact, direct assays for kaurene biosynthesis in the liquid endosperm of pea seeds yielded negative results. The activity in pea seeds increases to a maximum when the seeds are approximately half-maximum fresh weight (see Fig. 3-20) and the cotyledons nearly fill the seed cavity and then decreases as the seeds mature. Thus, so far as can be determined on the basis of studies to date, GA biosynthesis in developing seeds of *Pisum sativum* and *Marah macrocarpus* occurs in anatomically quite dissimilar parts of the seeds, but parts which, in both kinds of seeds, are most commonly associated with a nutritive function. Of course the incorporation of mevalonic acid into kaurene in cell-free enzyme extracts does not necessarily mean that these intermediates are converted to GAs in these tissues. However, the available data on extractable GA content in legume and *Marah macrocarpus* seeds correlate well with the published results concerning localization of kaurene-synthesizing enzymes.

Very little information is available on sites of GA synthesis in germinating seeds. Margaret Radley showed by extraction and bioassay in 1967 that GA-like substances are produced in the scutellum of barley embryos during the first 2 days of germination. On the third day, the activity of the scutellum ceases and the embryonic axis begins to produce GA.

Transport of GA

Most of the data concerning transport of GA in seed plants pertain to movement of applied (exogenous) GA. Information regarding movement of endogenous GA is largely indirect. A. J. McComb was among the first to investigate movement of exogenous GA. He applied [^{14}C]GA$_3$ to pea seedlings and reported in 1964 that translocation of applied GA was devoid of any unique features, GA transport apparently occurring in the phloem by the same mechanisms and according to similar patterns for other organic metabolites. Additional, more direct evidence for the transport of GAs in phloem comes from the findings that GAs occur in the sieve tube sap of plants.

T. Y. Chin and J. A. Lockhart concurred in 1965 with McComb's conclusions in their investigations of translocation of applied GA$_3$ in pinto bean seedlings. They concluded, as have other authors, that the movement of applied GA is related to carbohydrate transport within the plant.

Other, more recent investigations have shown that GAs are transported both in the phloem and the xylem, and there is experimental evidence from studies using [^{14}C]GA that interchange between phloem and xylem occurs. Thus, there appears to be a ready transport of GAs in the entire conductive (vascular) system of the plant, proceeding passively with the flow of assimilates or of water plus salts and some other organic compounds.

Most investigations of GA transport have employed exogenous excised sections of coleoptiles, stems, or petioles, exogenous radioactive GA, and the donor–receptor agar block technique which was described in the chapter on auxins. In general, movement has appeared to be nonpolar (in contrast to transport of endogenous auxin) and to occur at comparative rates of 5 to 25 mm per 12 hours. However, there have been reports of polar GA transport as well. When reported, polar transport has been described as being in the basipetal vector in stem, coleoptile, and subapical root sections. Perhaps the apparent polar movement, when observed, really is a movement from a source to a growth center (in accord with the concept of source-to-sink movement of metabolites in phloem generally), and really is not true polar movement.

Anatomical and Biophysical Basis of GA-Induced Growth

The most dramatic and most thoroughly studied effect of GAs is promotion of stem elongation, particularly in many genetic dwarf plants (Figs. 3-9 and 3-10) and the rosette stages of biennials and other plants that require vernalization and/or photoperiodic induction by long days to flower

(Fig. 7-20). Indeed, as previously indicated, it was the capacity of GAs to enhance elongation of rice stems that led to their discovery.

Since the early days of investigation of the effects of GAs on growth, it has been evident that these hormones can stimulate both cell division and cell enlargement. In the case of bean (*Phaseolus vulgaris*) seedling stems in situ, both parameters were stimulated by exogenous GA, but the largest effect was thought to be on cell elongation. Enhancement of cell enlargement has also been reported to be the major or exclusive response to exogenous GAs by *Tradescantia* stamen hairs, mesocarp parenchyma of parthenocarpic peaches, and intercalary meristems of excised internodal segments of *Avena sativa* stems.

In other cases, the major or primary effect of GAs is stimulation of cell division, as in subapical regions of the apical meristem of rosette plants such as *Hyoscyamus niger* and *Samolus parviflorus* (Fig. 3-24) and dwarf peas. Clearly the response of a given cell or tissue to divide or elongate depends upon age and stage of development. Younger cells generally can respond by dividing while older cells may respond only by enlargement.

Distribution of cell division

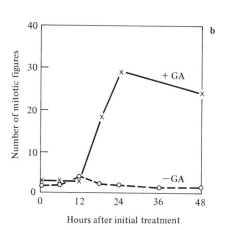

Hours after initial treatment

Figure 3-24. GA applications to rosette plants produce bolting by increasing transverse cell divisions in the tissues immediately below the apical meristem. Longitudinal sections through the shoot apex of *Hyoscyamus niger* illustrate where division increases following application of GA, **a**. (Each dot represents one mitotic figure in a 64-micron-thick section.) The numbers of mitoses in the subapical tissues of shoot apices of *Hyoscyamus niger*, with and without GA, are shown in **b**. (Redrawn, with permission, from Sachs et al., 1959.)

In general, the stem elongation response manifested by whole plants to treatment with exogenous GAs is correlated with effects of the hormones on both cell division and cell enlargement, the former effect of course preceding the latter in time (Fig. 3-24). As noted in Chapter 1, cell division alone does not result in growth, and so stimulation of cell division must be coupled with a normal or enhanced rate of cell elongation in order for there to be a growth response to exogenous GAs.

Interestingly, it appears that GAs are not essential for growth of roots. It has not been shown convincingly that GAs either generally stimulate or inhibit root growth or that GAs affect cell division or cell elongation in roots.

Although both auxins and GAs promote cell enlargement, it is not yet clear whether they do so in the same manner. Whereas the major effect of auxins is to increase cell wall plasticity (see Chapter 2), the effects of GAs are less well understood. Cleland and associates (1968) compared the effect of auxin and GA on the cell wall extensibility of cucumber hypocotyls, the growth of which is promoted by both hormones. Their results showed a consistently strong promotion of plasticity by IAA, but GA caused a variable plasticity response with values only slightly above control levels. They concluded, therefore, that GA had little or no effect on cell wall plasticity. However, in later investigations with *Avena* stem segments, Adams et al. (1975) found that GA markedly increased cell wall plasticity, in a manner resembling the action of auxin (Fig. 3-25). The plasticity values of GA-treated segments were, on the average, three times greater than those of control *Avena* stem segments. These authors were led to speculate that ". . . plant cell elongation might occur by a similar mechanism regardless of where it occurs and which factors control its occurrence."

Mechanism of Action of GA

It is well known that germinating seeds produce hydrolytic enzymes that digest the fat, carbohydrate, and protein reserves of the storage tissues. Investigations of the tissue source of these hydrolases and the control of their production and release have provided the most definitive information on the mechanism of action of GA. In particular, investigations of germinating barley grains (*Hordeum vulgare*) have been fruitful.

Early History

Haberlandt noted in 1890 that the aleurone layers of rye produce a substance (or substances) which causes liquefaction of the starchy portion of the endosperm and dissolution of the starch grains. Brown and Escombe confirmed this observation in 1898, using barley grains.

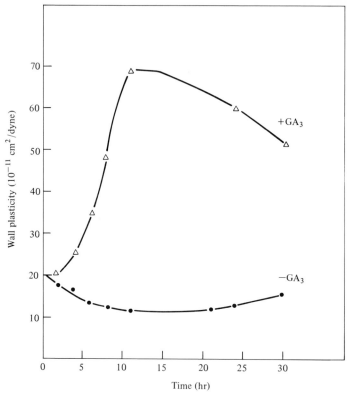

Figure 3-25. Effect of GA on cell wall plasticity in *Avena* stem segments. Segments were incubated in 0.1 *M* sucrose plus or minus 30 μM GA. (Redrawn with modifications, by permission, from Adams et al., 1975.)

In 1960, L. G. Paleg, in Australia, and H. Yomo, in Japan, showed independently that when gibberellic acid (GA₃) is added to endosperm portions of barley grains, it initiates the apparent production of certain amylolytic enzymes, including α-amylase, and the release of sugars as well as stimulating germination. D. E. Briggs extended this observation in 1963 to include several other hydrolytic enzymes (protease, phosphatase, and β-glucanase). Beginning in 1964, J. E. Varner and several associates at the MSU/DOE Plant Research Laboratory at Michigan State University, East Lansing, became leaders in this field.

Description of the Barley Experimental System

In the course of natural germination of a barley grain, the embryo is the source of endogenous GA. Margaret Radley showed in 1967 that GA-like substances are produced by the scutellum during the first 2 days of germination and thereafter by the embryo axis (Fig. 3-26). The GAs liberated by

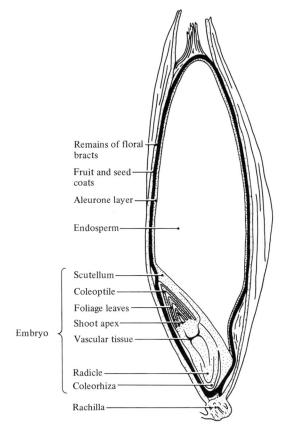

Figure 3-26. Longitudinal section of a barley grain. (Redrawn from *Control Mechanisms in Plant Development* by A. W. Galston and P. J. Davies, p. 108. Copyright © 1970 by Prentice-Hall, Inc. By permission of Prentice-Hall, Inc.)

Remains of floral bracts

Fruit and seed coats

Aleurone layer

Endosperm

Scutellum

Coleoptile

Foliage leaves

Shoot apex

Embryo

Vascular tissue

Radicle

Coleorhiza

Rachilla

the embryo diffuse across the endosperm to the aleurone cells, in which the hydrolytic enzymes are produced. Besides the embryo, the only other living cells in the grain are those of the aleurone, which is composed of several layers of cells constituting the outermost layer of endosperm. Hence the aleurone cells are a homogeneous collection of respiring, non-dividing cells which have the highly specialized function of producing and releasing some of those enzymes which are required to digest the starchy endosperm in preparation for its utilization by the growing embryo.

By 1961, investigators were utilizing embryo-less half-grains of barley, with which it could be shown that the increase in α-amylase activity is absolutely dependent upon added GA. Within a few more years, many investigators turned to investigations of isolated aleurone layers, which are separated from the starchy portion of the endosperm after the embryo-less half seeds have been imbibed for 3 days.

When isolated aleurone layers of barley are incubated in a solution containing GA, they produce and secrete several hydrolytic enzymes. GA-dependent de novo synthesis has been demonstrated for α-amylase, protease, β-1,3-glucanase, and ribonuclease. In addition, a GA-dependent release of ribonuclease and β-1,3-glucanase has been demonstrated. The increase in activity of at least one hydrolase, β-amylase, in the presence of GA is due to release of preformed enzyme and not to de novo synthesis.

Evidence that the increase in activity of at least four hydrolases induced by GA actually occurs as a result of de novo synthesis, rather than by activation of preformed enzyme, has been obtained in various ways. However, the most unequivocal proof comes from density labeling experiments, first performed by P. Filner and J. E. Varner in 1967. The procedure was to incubate barley aleurone layers in a solution containing GA_3 and either normal water or water containing the heavy isotope of oxygen ($H_2^{18}O$). Hydrolysis of stored proteins in the aleurone cells liberates amino acids for new protein synthesis. Hence, hydrolysis of stored proteins in the presence of $H_2^{18}O$ results in the liberation of amino acids labeled with the heavy isotope of oxygen:

$$\text{Protein } (n \text{ amino acid residues}) + n \text{ } H_2^{18}O \rightarrow$$
$$n \text{ } RCH-NH_2-C^{18}OOH \text{ and } RCH-NH_2-CO^{18}OH$$

Any proteins synthesized from the labeled amino acids will be heavier than proteins formed from amino acids released in the presence of ordinary $H_2^{16}O$; that is, the former will be "density-labeled" (Fig. 3-27). Upon

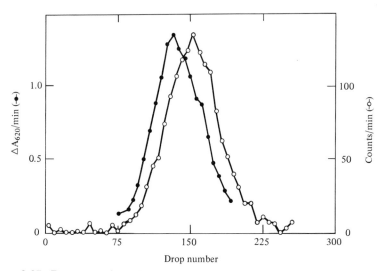

Drop number

Figure 3-27. Demonstration of the stimulation by GA_3 of de novo synthesis of α-amylase in barley aleurone layers by the density-labeling technique of Filner and Varner (1967). Isolated barley aleurone layers were incubated for 36 hours in the presence of GA_3 and either $H_2^{18}O$ or ordinary $H_2^{16}O$. The endogenous reserve proteins are hydrolyzed to amino acids which then are available for reuse in protein synthesis. If this protein turnover occurs in the presence of $H_2^{18}O$, the carbonyl conducting isopycnic equilibrium centrifugation (e.g., 40,000 rpm for 65 to 68 hours) in a cesium chloride gradient, it could be shown that the α-amylase appearing in GA_3-treated aleurone cells incubated with $H_2{}^{18}O$ was of approximately the theoretical 1% greater buoyant density than α-amylase from $H_2{}^{16}O$-incubated aleurone cells. These results proved that all the induced α-amylase activity was due to de novo synthesis of the enzyme.

In 1967 J. V. Jacobsen and J. E. Varner proved by the same procedures that protease also is synthesized de novo in response to GA. And P. A. Bennett and M. J. Chrispeels (1972) proved GA-induced de novo synthesis of ribonuclease and β-1,3-glucanase in barley aleurone cells. Incidentally, they used D_2O instead of $H_2{}^{18}O$.

Characteristics of the GA-Induced de Novo Synthesis of Hydrolases in Barley Aleurone Cells

Induction of de novo synthesis of α-amylase by GA in isolated aleurone layers is evident after a lag period of approximately 8 hours following administration of the hormone (Fig. 3-28). α-Amylase appears to be released from the aleurone cells as it is synthesized. But ribonuclease (Fig. 3-29) and β-1,3-glucanase first accumulate in the aleurone cells and then are released or secreted rapidly. In keeping with hormone responses generally, GA must be present continuously if the de novo synthesis of hydrolases is to be sustained (Fig. 3-30).

Synthesis of new RNA is essential to the GA-induction of de novo synthesis of hydrolases. Actinomycin D, an inhibitor of RNA synthesis, inhibits the synthesis and release of α-amylase if the inhibitor is presented during the first 7 to 8 hours after treatment. Thereafter the process is insensitive to actinomycin D. This is interpreted as indicating that all the necessary messenger RNA (mRNA) for α-amylase synthesis is formed during the initial (lag) period. Inhibitors of protein synthesis, such as cy-

---

oxygens of the peptide chain contain one-half the atom% ^{18}O present in the $H_2{}^{18}O$:

$$\text{Reserve protein} + H_2{}^{18}O \rightarrow (RCHNH_2CO^{18}OH \leftrightarrow RCHNH_2C^{18}OOH)$$

In the experiment results of which are illustrated above, marker α-amylase labeled with ^3H and nonradioactive but ^{18}O-labeled α-amylase were simultaneously subjected to isopycnic equilibrium centrifugation on a CsCl gradient. The ^3H-labeled marker enzyme was located by radioactivity, and ^{18}O-labeled enzyme was located by enzyme assay. The distribution of ^{18}O-labeled enzyme was displaced 15 drops in the direction of higher density, which represents a 1.1% increase in mass of the newly synthesized enzyme. Calculations showed that essentially all of the α-amylase formed in aleurone layers incubated in $H_2{}^{18}O$ and GA_3 arose by de novo synthesis from free amino acids derived from preexisting aleurone proteins. (Redrawn, with permission, from Filner and Varner, 1967.)

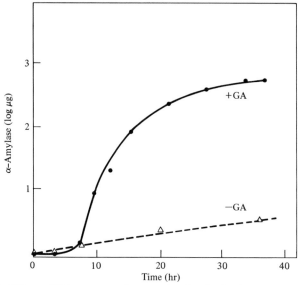

Figure 3-28. Effect of GA on α-amylase production by barley endosperm. Ten endosperm halves were incubated in buffer with 10^{-6} M GA$_3$ where indicated. (Redrawn, with permission, from Varner et al., 1965.)

cloheximide, also inhibit GA-induction of hydrolases, as expected. Interestingly, abscisic acid, a growth-inhibiting hormone (see Chapter 5), also inhibits GA-induced α-amylase synthesis (Fig. 3-31).

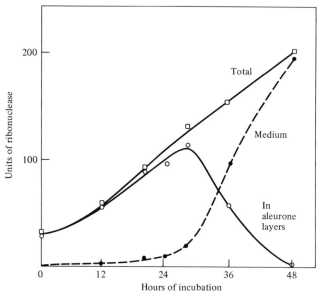

Figure 3-29. Time course of production and release of ribonuclease by 10 aleurone layers incubated with 5 mμM GA. (Redrawn, with permission, from Chrispeels and Varner, 1967a.)

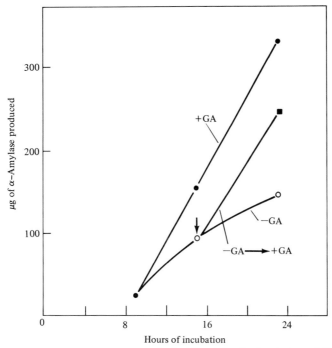

Figure 3-30. Effect of removing GA and of adding it back after being withheld for 6 hours on de novo synthesis of α-amylase in isolated aleurone layers of barley grains. (Redrawn, with permission, from Chrispeels and Varner, 1967b.)

Site and Mechanism of Action of GA in the Aleurone System

It is clear that GA causes derepression of the genes for the synthesis of α-amylase and some other hydrolases in barley aleurone layers. However, it is not clear from the evidence presented thus far whether the hormone operates directly at the transcriptional level (during mRNA synthesis) at the posttranscriptional level, or at the translational level (during protein synthesis). A close look at the events which occur during the lag period in GA induction of de novo synthesis of new enzymes has provided some important clues, however.

W. H. Evins and J. E. Varner in 1971 and 1972 reported on studies of the biochemical sequence of events during the lag period in the GA-induced synthesis of α-amylase in isolated aleurone layers. They found an increase in polyribosome formation and an increased synthesis of ribosomes and endoplasmic reticulum membranes. All these effects begin within 2 to 4 hours after application of GA. They showed that the polyribosomes responsible for new enzyme synthesis were those polyribosomes produced in the presence of the hormone, and that those polyribo-

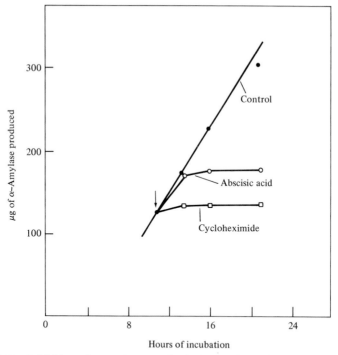

Figure 3-31. Inhibition of α-amylase synthesis by abscisic acid and cycloheximide. Aleurone layers were incubated in 0.1 μM GA for 11 hours. At this time abscisic acid (5 μM) or cycloheximide (10 μg/ml) was added and α-amylase synthesis was measured 2.5, 5, and 10 hours later. (Redrawn, with permission, from Chrispeels and Varner, 1967b.)

somes were attached to the endoplasmic reticulum. These observations led Evins and Varner to conclude that the GA-stimulated increases in the number of monoribosomes and the percentage of polyribosomes probably are prerequisite for the hormone induction of protein synthesis. Incidentally, abscisic acid (Chapter 5), which already has been shown to counteract the effect of GA in the barley aleurone system, inhibits RNA synthesis, which in turn results in inhibition of polyribosome formation. The results of Evins and Varner are consistent with the notion that GA evokes selective mRNA synthesis, which is correlated with enhanced polyribosome formation, and de novo synthesis of particular hydrolases in the aleurone system. Regarding the effect of GA on membrane proliferation before the increase of hydrolase activity, there is much evidence for the participation of rough endoplasmic reticulum in the synthesis and exportation of secreted proteins in eukaryotic cells.

K. D. Johnson and H. Kende reported in 1971 on some early biochemical effects that occur in barley aleurone layers during the lag period in the response to exogenous GA. They found that the earliest detectable changes after GA treatment were in the activities of two enzymes involved in lecithin (a common component of membranes) biosynthesis, namely phosphoryl choline glyceride transferase and phosphoryl choline cytidyl transferase. Both of these enzymes increase in activity about threefold during the first 2 hours after treatment of aleurone layers with GA. Accompanying, perhaps even preceding, changes in the two enzyme activities, they observed an increased incorporation of ^{32}P into cytidyl triphosphate (CTP), and several hours later an increased incorporation of labeled choline into microsomal lipid and of ^{32}P into phospholipid. The latter changes all were detectable within 4 hours after treatment. On the basis of these results, Johnson and Kende proposed a posttranscriptional model for GA action, according to which synthesis of endoplasmic reticulum is initiated within the first few hours after treatment, completion of that activity occurs some 4 to 8 hours later, and finally there is secretion of hydrolytic enzymes the synthesis of which occurs on mRNA templates already present in the aleurone before GA treatment. The latter point is not supported by direct evidence, and indeed conflicts obviously with much data signifying the dependence of GA action on new RNA synthesis.

Recently (1972), J. A. Zwar and J. V. Jacobsen showed that GA stimulates selective incorporation of labeled precursor into polydisperse (5 to 14 S) RNA. Very recent reports that GA increases the synthesis of poly(A)-containing RNA provides further support for the view that the hormone acts at the transcriptional level, since poly(A) segments are characteristic of eukaryotic mRNAs. The enhancement of poly(A)-RNA synthesis begins within 3 to 4 hours after presentation of GA and attains a maximum at 10 to 12 hours, which is just about the time α-amylase first becomes detectable.

Still more recently, in 1976, T. J. V. Higgins and associates provided an even more direct link between GA-stimulated de novo enzyme synthesis and appearance of the complementary mRNA. They isolated total RNA and poly(A)-containing RNA from aleurone layers and assayed for α-amylase in a cell-free system from wheat germ, and analyzed the products of translation by immunoprecipitation and polyacrylamide gel electrophoresis. They demonstrated convincingly that the level of translatable α-amylase mRNA increased in GA-treated tissue in parallel with the increased rate of enzyme synthesis (Fig. 3-32). These results provide still more evidence that GA acts to induce selective mRNA and de novo enzyme synthesis in aleurone cells.

Action of GAs in Other Systems

The barley aleurone system is an excellent one for investigating the site and mechanism of action of GAs, and much has been learned with that system. But it must be considered to what extent analogous action is observed in other systems if we are to eventually fully understand the role of GAs in the regulation of growth and development.

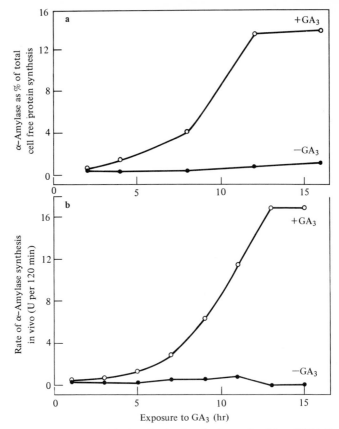

Figure 3-32. The increase with time in the level of translatable mRNA for α-amylase and the increase in the rate of synthesis of the enzyme in vivo in response to
GA₃ treatment. **a,** α-Amylase synthesized by the cell-free system was estimated
from densitometric scans of autoradiographs of gels which had been in contact
with X-ray film for different times. The gels were scanned with a microdensito-
meter and the contribution made by the α-amylase peak to total protein synthesized was calculated by weight. **b,** The rate of α-amylase synthesis in vivo was
calculated as the increment in enzyme level (units of α-amylase) during successive
2-hour intervals following hormone treatment. Total α-amylase activity was assayed, using replicates of 10 aleurone layers. U, Units of α-amylase. (Redrawn,
with permission, from Higgins et al., 1976.)

One very interesting investigation with another system was conducted
by M. M. Johri and J. E. Varner in 1968. They studied enhancement of
RNA synthesis in isolated nuclei of dwarf pea seedlings by GA. They detected marked increases in RNA synthesis when GA₃ was present
throughout the isolation procedures (Fig. 3-33). GA₃ is also active in promoting stem elongation in dwarf peas. Hence it was of interest to see what
effect GA₈, which does not promote growth of dwarf peas, would have on

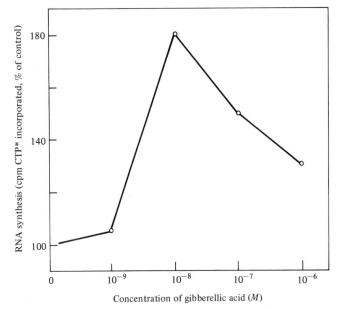

Figure 3-33. The application of GA$_3$ during the isolation of nuclei from pea seedlings results in an increased RNA synthesis in the nuclei, as indicated by the incorporation of [^{14}C]cytidine triphosphate (CTP). (Drawn from data of Johri and Varner, 1968.)

the system. The result was as predicted; GA$_8$ did not evoke enhanced RNA synthesis by the isolated dwarf pea nuclei.

L. G. Paleg and associates have suggested since 1972, on the basis of results with artificial model membrane systems, that GAs act by regulating membrane permeability. It will be interesting to follow future investigations designed to test that hypothesis.

Conclusions Regarding GA Action

On the basis of collective data now available, it may confidently be concluded that in the barley aleurone system GA evokes selectively the synthesis of particular molecular species of mRNA which in turn leads to de novo synthesis of certain hydrolytic enzymes. Left unanswered unequivocally at this stage of our knowledge are two important questions. One is whether the action of GA just described alone accounts exclusively for the fundamental mechanism of action of GA in the aleurone system. The other question is whether the action of GA in stimulating de novo synthesis of particular enzymes as described for the barley aleurone system is universal. That is, is the mechanism of GA action in the barley aleurone system directly indicative of the mechanism by which GA hormones participate generally in the regulation of the growth and development of higher plants? Obviously, the answers to these questions must await additional research.

References

Adams, P. A., M. J. Montague, M. Tepfer, D. L. Rayle, H. Ikuma, and P. B. Kaufman. 1975. Effect of gibberellic acid on the plasticity and elasticity of *Avena* stem segments. *Plant Physiol.* **56**: 757–760.

Barendse, G. W. M. 1971. Formation of bound gibberellins in *Pharbitis nil. Planta* **99**: 290–301.

Barendse, G. W. M. 1975. Biosynthesis, metabolism, transport and distribution of gibberellins. In: Krishnamoorthy, H. N., ed. *Gibberellins and Plant Growth.* Wiley Eastern Limited, New Delhi. Pp. 65–89.

Barendse, G. W. M., H. Kende, and A. Lang. 1968. Fate of radioactive gibberellin A_1 in maturing and germinating seeds of peas and Japanese morning glory. *Plant Physiol.* **43**: 815–822.

Barendse, G. W. M. and A. Lang. 1972. Comparison of endogenous gibberellins and of the fate of applied radioactive gibberellin A_1 in a normal and a dwarf strain of Japanese morning glory. *Plant Physiol.* **49**: 836–841.

Bennett, P. A. and M. J. Chrispeels. 1972. *De novo* synthesis of ribonuclease and β-1,3-glucanase by aleurone cells of barley. *Plant Physiol.* **49**: 445–447.

Brian, P. W., G. W. Elson, H. G. Hemming, and M. Radley. 1954. The plant-growth promoting properties of gibberellic acid, a metabolic product of the fungus *Gibberella fujikuroi. J. Sci. Food. Agr.* **5**: 602–612.

Chrispeels, M. J. and J. E. Varner. 1967a. Gibberellic acid-enhanced synthesis and release of α-amylase and ribonuclease by isolated barley aleurone layers. *Plant Physiol.* **42**: 398–406.

Chrispeels, M. J. and J. E. Varner. 1967b. Hormonal control of enzyme synthesis: on the mode of action of gibberellic acid and abscisin in aleurone layers of barley. *Plant Physiol.* **42**: 1008–1016.

Cleland, R., M. L. Thompson, D. L. Rayle, and W. K. Purves. 1968. Differences in effects of gibberellins and auxins on wall extensibility of cucumber hypocotyls. *Nature (London)* **219**: 510–511.

Coolbaugh, R. C. and R. Hamilton. 1976. Inhibition of *ent*-kaurene oxidation and growth by α-cyclopropyl-α-(*p*-methoxyphenyl)-5-pyrimidine methyl alcohol. *Plant Physiol.* **57**: 245–248.

Coolbaugh, R. C. and T. C. Moore. 1969. Apparent changes in rate of kaurene biosynthesis during the development of pea seeds. *Plant Physiol.* **44**: 1364–1367.

Coolbaugh, R. C. and T. C. Moore. 1971. Localization of enzymes catalysing kaurene biosynthesis in immature pea seeds. *Phytochemistry* **10**: 2395–2400.

Coolbaugh, R. C., T. C. Moore, S. A. Barlow, and P. R. Ecklund. 1973. Biosynthesis of *ent*-kaurene in cell-free extracts of *Pisum sativum* shoot tips. *Phytochemistry* **12**: 1613–1618.

Ecklund, P. R. and T. C. Moore. 1968. Quantitative changes in gibberellin and RNA correlated with senescence of the shoot apex in the 'Alaska' pea. *Am. J. Bot.* **55**: 494–503.

Ecklund, P. R. and T. C. Moore. 1974. Correlations of growth rate and de-etiolation with rate of *ent*-kaurene biosynthesis in pea (*Pisum sativum* L.). *Plant Physiol.* **53**: 5–10.

Evins, W. H. and J. E. Varner. 1972. Hormonal control of polyribosome formation in barley aleurone layers. *Plant Physiol.* **49**: 348–352.

Filner, P. and J. E. Varner. 1967. A test for *de novo* synthesis of enzymes: Density labeling with $H_2^{18}O$ of barley α-amylase induced by gibberellic acid. *Proc. Natl. Acad. Sci. U.S.A.* **58**: 1520–1526.

Galston, A. W. 1961. *The Life of the Green Plant.* Prentice-Hall, Englewood Cliffs, New Jersey.

Galston, A. W. and P. J. Davies. 1970. *Control Mechanisms in Plant Development.* Prentice-Hall, Englewood Cliffs, New Jersey.

Graebe, J. E. 1968. Biosynthesis of kaurene, squalene and phytoene from mevalonate-2-^{14}C in a cell-free system from pea fruits. *Phytochemistry* **7**: 2003–2020.

Graebe, J. E., D. H. Bowen, and J. MacMillan. 1972. The conversion of mevalonic acid into gibberellin A_{12}-aldehyde in a cell-free system from *Cucurbita pepo. Planta* **102**: 261–271.

Graebe, J. E., D. T. Dennis, C. D. Upper, and C. A. West. 1965. Biosynthesis of gibberellins. I. The biosynthesis of (−)-kaurene, (−)-kaurene-19-ol, and *trans*-geranylgeraniol in endosperm nucellus of *Echinocystis macrocarpa* Greene. *J. Biol. Chem.* **240**: 1847–1854.

Hasson, E. P. and C. A. West, 1976a. Properties of the system for the mixed function oxidation of kaurene and kaurene derivatives in microsomes of the immature seed of *Marah macrocarpus*. Cofactor requirements. *Plant Physiol.* **58**: 473–478.

Hasson, E. P. and C. A. West. 1976b. Properties of the system for the mixed function oxidation of kaurene and kaurene derivatives in microsomes of the immature seed of *Marah macrocarpus*. Electron transfer components. *Plant Physiol.* **58**: 479–484.

Hedden, P., J. MacMillan, and B. O. Phinney. 1978. The metabolism of the gibberellins. *Ann. Rev. Plant Physiol.* **29**: 149–192.

Hedden, P. and B. O. Phinney. 1976. The dwarf-5 mutant of *Zea mays:* a genetic lesion controlling the cyclization step (B activity) in kaurene biosynthesis. In: Pilet P.-E., ed. *Collected Abstracts of the Paper-Demonstrations of the 9th International Conference on Plant Growth Substances.* P. 136 (Abstract).

Higgins, T. J. V., J. A. Zwar, and J. V. Jacobsen. 1976. Gibberellic acid enhances the level of translatable mRNA for α-amylase in barley aleurone layers. *Nature (London)* **260**: 166–169.

Iwahori, S., R. H. Weaver, and R. M. Pool. 1968. Gibberellin-like activity in berries of seeded and seedless Tokay grapes. *Plant Physiol.* **43**: 333–337.

Jackson, D. I. 1968. Gibberellin and the growth of peach and apricot fruits. *Austral. J. Biol. Sci.* **21**: 209–215.

Jackson, D. I. and B. G. Coombe. 1966. Gibberellin-like substances in the developing apricot fruit. *Science* **154**: 277–278.

Jacobsen, J. V. 1977. Regulation of ribonucleic acid metabolism by plant hormones. *Ann. Rev. Plant Physiol.* **28**: 537–564.

Jacobsen, J. V. and J. E. Varner. 1967. Gibberellic acid-induced synthesis of protease by isolated aleurone layers of barley. *Plant Physiol.* **42**: 1596–1600.

Johnson, K. D. and H. Kende. 1971. Hormonal control of lecithin synthesis in barley aleurone cells: regulation of the CDP-choline pathway by gibberellin. *Proc. Natl. Sci. U.S.A.* **68**: 2674–2677.

Johri, M. M. and J. E. Varner. 1968. Enhancement of RNA synthesis of isolated pea nuclei by gibberellic acid. *Proc. Natl. Acad. Sci. U.S.A.* **59**: 269–276.

Jones. R. L. 1973. Gibberellins: their physiological role. *Ann. Rev. Plant Physiol.* **24**: 571–598.

Jones, R. L. and A. Lang. 1968. Extractable and diffusible gibberellins from light- and dark-grown pea seedlings. *Plant Physiol.* **43**: 629–634.

Jones, R. L. and I. D. J. Phillips. 1966. Organs of gibberellin synthesis in light-grown sunflower plants. *Plant Physiol.* **41**: 1381–1386.

Kende, H. 1967. Preparation of radioactive gibberellin A_1 and its metabolism in dwarf peas. *Plant Physiol.* **42**: 1612–1618.

Kende, H. and A. Lang. 1964. Gibberellins and light inhibition of stem growth in peas. *Plant Physiol.* **39**: 435–440.

Knotz, J., R. C. Coolbaugh, and C. A. West. 1977. Regulation of the biosynthesis of *ent*-kaurene from mevalonate in the endosperm of immature *Marah macrocarpus* seeds by adenylate energy charge. *Plant Physiol.* **60**: 81–85.

Köhler, D., 1966. Die Abhängigkeit der Gibberellinproduktion von Normalerbsen vom Phytochrom-system. *Planta* **69**: 27–33.

Krishnamoorthy, H. N., ed. 1975. *Gibberellins and Plant Growth*. Wiley Eastern Limited, New Delhi.

Lang, A. 1970. Gibberellins: structure and metabolism. *Ann. Rev. Plant Physiol.* **21**: 537–570.

Leopold, A. C. and P. E. Kriedemann. 1975. *Plant Growth and Development*. 2nd ed. McGraw-Hill Book Company, New York.

Liu, P. B. W. and J. B. Loy. 1976. Action of gibberellic acid on cell proliferation in the subapical shoot meristem of watermelon seedlings. *Am. J. Bot.* **63**: 700–704.

Lockhart, J. A. 1957. Studies on the organ of production of the natural gibberellin factor in higher plants. *Plant Physiol.* **32**: 204–207.

MacMillan, J. 1971. Diterpenes—the gibberellins. In: Goodwin, T. W., ed. *Aspects of Terpenoid Chemistry and Biochemistry*. Academic Press, London. Pp. 153–180.

MacMillan, J. 1977. Some aspects of gibberellin metabolism in higher plants. In: Pilet, P. E., ed. *Plant Growth Regulation*. Springer-Verlag, New York. Pp. 129–138.

McComb, A. J. 1964. The stability and movement of gibberellic acid in pea seedlings. *Ann. Bot. (N.S.)* **28**: 669–687.

Milborrow, B. V. 1974. Biosynthesis of abscisic acid by a cell-free system. *Phytochemistry* **13**: 131–136.

Mitchell, J. W., D. P. Skaggs, and W. P. Anderson. 1951. Plant growth-stimulating hormones in immature bean seeds. *Science* **114**: 159–161.

Moore, T. C. 1967. Gibberellin relationships in the 'Alaska' pea (*Pisum sativum*). *Am. J. Bot.* **54**: 262–269.

Moore, T. C., S. A. Barlow, and R. C. Coolbaugh. 1972. Participation of noncatalytic 'carrier' protein in the metabolism of kaurene in cell-free extracts of pea seeds. *Phytochemistry* **11**: 3225–3233.

Moore, T. C. and P. R. Ecklund. 1974. Biosynthesis of *ent*-kaurene in cell-free extracts of pea shoots. In: *Plant Growth Substances 1973*. Hirokawa Publishing Company, Tokyo. Pp. 252–259.

Moore, T. C. and P. R. Ecklund. 1975. Role of gibberellins in the development of fruits and seeds. In: Krishnamoorthy, H. N., ed. *Gibberellins and Plant Growth*. Wiley Eastern Limited, New Delhi. Pp. 145–182.

Murakami, Y. 1961. Paper-chromatographic studies on change in gibberellins during seed development and germination in *Pharbitis nil. Bot. Mag. (Tokyo)* **74**: 241–247.

Murphy, P. J. and C. A. West. 1969. The role of mixed function oxidases in kaurene metabolism in *Echinocystis macrocarpa* Greene endosperm. *Arch. Biochem. Biophys.* **133**: 395–407.

Ogawa, Y. 1964. Changes in the amount of gibberellin-like substances in the seedling of *Pharbitis nil* with special reference to expansion of cotyledon. *Plant Cell Physiol.* **5**: 11–20.

Paleg, L. G. 1965. Physiological effects of gibberellins. *Ann. Rev. Plant Physiol.* **16**: 291–322.

Phillips. I. D. J. 1971. *Introduction to the Biochemistry and Physiology of Plant Growth Hormones*. McGraw-Hill Book Company, New York.

Phinney, B. O., C. A. West, M. Ritzel, and P. M. Neely. 1957. Evidence for "gibberellin-like" substances from flowering plants. *Proc. Natl. Acad. Sci. U.S.A.* **43**: 398–404.

Radley, M. 1956. Occurrence of substances similar to gibberellic acid in higher plants. *Nature (London)* **178**: 1070–1071.

Radley, M. 1958. The distribution of substances similar to gibberellic acid in higher plants. *Ann. Bot. (N.S.)* **22**: 297–307.

Reeve, D. R. and A. Crozier. 1974. An assessment of gibberellin structure-activity relationships. *J. Exp. Bot.* **25**: 431–445.

Reid, D. M., M. S. Tuing, R. C. Durley, and I. D. Railton. 1972. Red-light-enhanced conversion of tritiated gibberellin A_9 into other gibberellin-like substances in homogenates of etiolated barley leaves. *Planta* **108**: 67–75.

Robinson, D. R. and C. A. West. 1970. Biosynthesis of cyclic diterpenes in extracts from seedlings of *Ricinus communis* L. I. Identification of diterpene hydrocarbons formed from mevalonate. *Biochemistry* **9**: 70–79.

Runeckles, V. C., E. Sondheimer, and D. C. Walton, eds. 1974. *The Chemistry and Biochemistry of Plant Hormones*. Vol. 7. *Recent Advances in Phytochemistry*. Academic Press, New York.

Sachs, R. M. 1965. Stem elongation. *Ann. Rev. Plant Physiol.* **16**: 73–96.

Sachs, R. M., C. F. Bretz, and A. Lang. 1959. Shoot histogenesis: the early effects of gibberellin upon stem elongation in two rosette plants. *Am. J. Bot.* **46**: 376–384.

Sembdner, G., J. Weiland, O. Aurich, and K. Schreiber. 1968. Isolation, structure and metabolism of a gibberellin glucoside. *S. C. I. Monograph No. 31:* 70–86.

Sitton, D., A. Richmond, and Y. Vaadia. 1967. On the synthesis of gibberellins in roots. *Phytochemistry* **6**: 1101–1105.

Stodola, F. H., K. B. Raper, D. I. Fennell, H. F. Conway, V. E. Sohns, C. T. Langford, and R. W. Jackson. 1955. The microbiological production of gibberellins A and X. *Arch. Biochem. Biophys.* **54**: 240–245.

Stowe, B. B. and T. Yamaki. 1957. The history and physiological action of the gibberellins. *Ann. Rev. Plant Physiol.* **8**: 181–216.

Stuart, D. A. and R. L. Jones. 1977. Roles of extensibility and turgor in gibberellin- and dark-stimulated growth. *Plant Physiol.* **59**: 61–68.

Thimann, K. V. 1974. Fifty years of plant hormone research. *Plant Physiol.* **54**: 450–453.

Thimann, K. V. 1977. *Hormone Action in the Whole Life of Plants.* University of Massachusetts Press, Amherst.

Varner, J. E. and D. T. Ho. 1976. Hormones. In: Bonner, J. and J. E. Varner, eds. *Plant Biochemistry.* 3rd ed. Academic Press, New York. Pp. 713–770.

Varner, J. E., G. Ram Chandra, and M. J. Chrispeels. 1965. Gibberellic acid-controlled synthesis of α-amylase in barley endosperm. *J. Cell. Comp. Physiol.* **66,** Suppl. 1: 55–68.

Wareing, P. F. and A. K. Seth. 1967. Ageing and senescence in the whole plant. *Symp. Soc. Exp. Biol.* **21**: 543–558.

West, C. A. 1973. Biosynthesis of gibberellins. In: Milborrow, B. V., ed. *Biosynthesis and its Control in Plants.* Academic Press, London. Pp. 143–169.

West, C. A. and B. O. Phinney. 1956. Properties of gibberellin-like factors from extracts of higher plants. *Plant Physiol.* **31**: (Suppl.) XX (Abstract).

Wiltbank, W. J. and A. H. Krezdorn. 1969. Determination of gibberellins in ovaries and young fruits of navel oranges and their correlation with fruit growth. *J. Am. Soc. Hort. Sci.* **94**: 195–201.

Yokota, T., N. Murofushi, N. Takahashi, and M. Katsumi. 1971. Biological activities of gibberellins and their glucosides in *Pharbitis nil. Phytochemistry* **10**: 2943–2949.

Zwar, J. A. and J. V. Jacobsen. 1972. A correlation between a ribonucleic acid fraction selectively labeled in the presence of gibberellic acid and amylase synthesis in barley aleurone layers. *Plant Physiol.* **49**: 1000–1006.

Chapter 4

Cytokinins

History of Discovery

The definitive discovery of the cytokinins occurred in 1955 when C. O. Miller, Folke Skoog, M. H. von Saltza, and F. M. Strong, working in the laboratories of Skoog and Strong at the University of Wisconsin, isolated a substance called "kinetin" (6-furfurylaminopurine, $C_{10}H_9N_5O$) (Fig. 4-1) from an autoclaved sample of herring sperm DNA and demonstrated it to be very active in promoting mitosis and cell division in tobacco callus tissue in vitro.

The idea of specific substances required for cell division occurring in plants actually dates back to the German physiologist J. Wiesner who, in 1892, proposed that initiation of cell division is evoked by endogenous factors, indeed a proper balance among endogenous factors. Somewhat later, Haberlandt reported in 1913 that diffusates from phloem tissue could induce cell division in parenchyma tissue of potato tubers. Repeating and extending some of Haberlandt's experiments, Jablonski and Skoog (1954) showed that a piece of vascular tissue cultured on top of tobacco pith tissue could cause division of the pith cells, which would not otherwise divide. These observations led to a search for pure substances which could induce cell division in a manner similar to that of the unknown substance or substances in vascular tissue.

Hence the ultimate discovery of the cytokinins was a direct outcome of tissue culture investigations which were carried on since circa 1941 in the laboratory of Folke Skoog at the University of Wisconsin. Skoog and his associates have utilized mostly aseptically excised portions of the nondividing mature pith from stems of well-developed tobacco (*Nicotiana tabacum* cv. Wisconsin No. 38) plants.

Figure 4-1. Structure of kinetin [6-(furfurylamino)-purine], the first cytokinin to be discovered.

Some important features of this tobacco pith tissue are as follows: (1) the tissue is mature and the parenchyma cells comprising it normally do not divide if efforts are made to culture the tissue in vitro on a simple culture medium containing no hormones; (2) both exogenous auxin and cytokinin are required for growth of the pith explants, and at appropriate concentrations of the two hormones the tissue grows into an amorphous, relatively undifferentiated callus; (3) if only auxin (and not cytokinin) is supplied, the pith tissue exhibits pronounced cell enlargement entirely unaccompanied by cell division; and (4) cytokinin is inactive in promoting cell division in the absence of auxin.

In passing it should be noted that the behavior of tobacco pith in tissue culture is not exhibited by all tissues. In fact a great many tissues are like tobacco pith in that they do require a cytokinin or auxin or both for growth on a tissue culture medium. But there are some examples of tissues which so far have not been grown on defined media successfully at all. And, at the other extreme, there are many examples of tissues, mostly meristematic tissues, which will grow very well on a relatively simple tissue culture medium containing only a few inorganic salts, a carbon-energy source, and certain vitamins—and no exogenous auxin or cytokinin. Meristematic and tumor tissues—such as Crown Gall tumor induced by *Agrobacterium tumefaciens*—exhibit this behavior. Evidently the cytokinin-synthesizing system is operative in the latter types of tissues, whereas it is turned off or repressed during the normal course of differentiation into a nondividing tissue such as pith.

Terminology

The name "kinetin" was assigned to 6-furfurylaminopurine by Miller et al. (1955) when they first reported on the isolation of the substance from DNA. The name of the substance obviously was based on its property of causing cell division or "cytokinesis" in tobacco pith tissue. Later, in 1956, the Wisconsin group proposed the generic term "kinin" for the group of synthetic and naturally occurring substances which resemble kinetin in inducing cell division in certain excised tissues in the presence of exogenous auxin.

Rather quickly, however, confusion began to develop over terminology, for animal physiologists at about the same time proposed the term "kinin" to designate a class of polypeptides (compounds of quite different nature than the plant hormones) which stimulate smooth muscle contraction and nerve endings in the manner of various stings and venoms. Hence, in efforts to avoid confusion, plant physiologists coined new terms such as "phytokinin," "cytonin," and "cytokinin" to replace "kinin" in the plant literature. Finally, after consultation with other investigators in the field, Skoog and his associates proposed that "cytokinin" be used to designate all compounds which promote cell division and exert other growth regulatory functions in the same manner as kinetin, and this term now is generally adopted. The term "kinetin," of course, is reserved for one specific cytokinin, namely 6-furfurylaminopurine.

Isolation of Kinetin and the Search for Other Naturally Occurring Cytokinins

Kinetin was first isolated from an autoclaved sample of DNA from herring sperm. Lest this be regarded incorrectly as a mere fortuitous discovery, it should be noted that Skoog and his associates had previously found that adenine had slight cell division-promoting activity in tobacco pith and they speculated that nucleic acids might be involved in the biological activity of the then unknown stimulus. In fact, a variety of purine-containing substances was tested, and among them an old sample of herring sperm DNA was found to be particularly effective in promoting growth. One can only imagine the surprise when Skoog et al. repeated the experiments which resulted in the discovery of kinetin with fresh samples of DNA and found that it was inactive.

It turns out that kinetin is an artifact, at least in terms of the high concentrations present in the old DNA from which it first was isolated. Only old or heated DNA preparations yielded substantial quantities of kinetin. Kinetin has the empirical formula of a dehydrated deoxyadenosine. It arises spontaneously in DNA preparations (as during autoclaving), presumably from deoxyadenosine. Apparently, during the transformation, the pentose derivative (i.e., the furfuryl constituent) migrates from the 9-position to the 6-position of the adenine ring, as illustrated below:

Figure 4-2. Structures of some synthetic cyto-
kinins. (Redrawn from *Plant Growth and De-
velopment,* 2nd ed., by A. C. Leopold and P. E.
Kriedemann. Copyright © 1975 by McGraw-
Hill Book Company. Used with permission of
McGraw-Hill Book Company.)

Kinetin

Benzyladenine (BA)

Tetrahydropyranylbenzyladenine (PBA)

Ethoxyethyladenine

Hall and deRopp (1955) demonstrated that kinetin could be produced simply by autoclaving a mixture of adenine and furfuryl alcohol. Most investigators now agree that kinetin is thus an artifact, although there has been
some speculation that kinetin may also be formed in vivo.

Artifact or not, still the biological activity of kinetin was real, and the
substance had activity such as had been expected of the elusive cell division-promoting hormone. Moreover, investigations of the biological activity of dozens of compounds resembling kinetin chemically showed that
kinetin-like biological activity was fairly well restricted to N^6-substituted
adenine derivatives, suggesting that perhaps some compound or compounds closely resembling kinetin was present naturally in plants. Indeed,
already by 1970, it was estimated that at least 100 potent cytokinins, all

N^6-substituted adenine compounds, had been synthesized (Fig. 4-2). Thus the search for naturally occurring kinetin-like compounds (i.e., cytokinins) intensified.

Discovery of Natural Cytokinins

The first isolation of a naturally occurring cytokinin should be credited to both D. S. Letham of the Fruit Research Division, DSIR, Auckland, New Zealand, and C. O. Miller of the University of Indiana, circa 1963. In 1963, D. S. Letham presented a report at a symposium in which he announced the isolation in crystalline form of a natural cytokinin from immature corn (*Zea mays*) kernels, which he called "zeatin." In a longer subsequent paper Letham et al. (1964) reported on many of the chemical properties of zeatin and postulated as the most probable structure 6-(4-hydroxy-3-methyl-*trans*-2-butenylamino)purine (Fig. 4-3), a structure promptly confirmed by Shaw and Wilson (1964). Interestingly, C. O. Miller (1961b) earlier had detected a kinetin-like substance in immature corn kernels, although he did not fully characterize the active compound. Later reports from Miller's laboratory and from the collaboration of Miller and Letham (1965) showed conclusively that the substance isolated by Miller also was zeatin. Zeatin, and its riboside and ribonucleotide have since been reported from many sources. As a matter of fact, Letham (1968) identified zeatin riboside as the major active cell division-promoting substance in coconut milk, which for some 27 years had been known to have growth-promoting properties. In the years following the discovery of zeatin, several other cytokinins were isolated from various sources and were characterized chemically (Fig. 4-3). All the naturally occurring cytokinins are considered to be isopentenyl adenine derivatives. N^6-(Δ^2-isopentenyl)adenosine (i^6A) occurs in all tRNA preparations that have been examined. In fact, i^6A is the only cytokinin that has been found in tRNA from animals. Four cytokinin-active ribonucleosides—i^6A, io^6A, and their 2-methylthio derivatives, ms^2i^6A and ms^2io^6A (Fig. 4-3)—have been identified from plant tRNA, although ms^2i^6A typically is present in only trace amounts if at all. Aside from glycosylation and methylthiolation of the purine ring, the important differences obviously are in the side chains. The free bases, illustrating these differences in side chain, are tabulated in Table 4-1.

By 1967, the cytokinin-containing extracts had been prepared from approximately 40 species of higher plants, and it is now confidently assumed that the cytokinins are ubiquitous among seed plants and perhaps throughout the plant kingdom. Indeed cytokinins have been reported to occur also, specifically in the tRNA, of numerous animals and microorganisms, indicating a high probability of occurrence, in fact, in all organisms.

Zeatin (io[6] Ade)

Dihydrozeatin

N[6]-(Δ[2]-Isopentenyl) adenine (i[6] Ade)

cis-Zeatin (cis-io[6] Ade)

2-Hydroxyzeatin

6-(3, 4-Dihydroxy-3-methylbutylamino) purine

6-(2, 3, 4-Trihydroxy-3-methylbutylamino) purine

N[6]-(2-Hydroxybenzyl) adenosine

Figure 4-3. Structures of some naturally occurring cytokinins. Illustrations include some free bases, ribosides, and ribotides.

Ribosylzeatin (io⁶A)

Ribosylzeatin-5′-monophosphate

N⁶-(Δ²-Isopentenyl) adenosine (i⁶A)

cis-Ribosylzeatin (io⁶A)

2-Methylthio-N⁶-(Δ²-isopentenyl) adenosine (ms²i⁶A)

2-Methylthio-cis-ribosylzeatin (ms²io⁶A)

Figure 4-3 (continued).

Table 4-1. Free Bases of Natural Cytokinins[a]

Purine nucleus with substituents NH—R_1 and R_2.

R_1	R_2	Chemical name	Common name or abbreviation
—CH_2—CH=$C(CH_3)$—CH_2OH	H	6-(4-Hydroxy-3-methyl-*trans*-2-butenyl)aminopurine	*trans*-Zeatin
—CH_2—CH=$C(CH_2OH)$—CH_3	H	6-(4-Hydroxy-3-methyl-*cis*-2-butenyl)aminopurine	*cis*-Zeatin
—CH_2—CH_2—$C(H)(CH_2OH)$—CH_3	H	6-(4-Hydroxy-3-methylbutyl)aminopurine	Dihydrozeatin

Structure	Name		Abbreviation
$-CH_2-CH=C(CH_3)CH_3$	H	6-(3-Methyl-2-butenyl)aminopurine	IPA
$-CH_2-CH=C(CH_3)CH_2OH$	CH_3S-	6-(4-Hydroxy-3-methyl-2-butenyl)2-methylthioaminopurine	CH$_3$S-Zeatin (*cis* or *trans*)
$-CH_2-CH=C(CH_3)CH_2OH$	CH_3S-	6-(4-Hydroxy-3-methyl-2-butenyl)2-methylthioaminopurine	CH$_3$S-Zeatin (*cis* or *trans*)
$-CH_2-CH=C(CH_3)CH_3$	CH_3S-	6-(3-Methyl-2-butenyl)2-methylthioaminopurine	CH$_3$S-IPA
$O=C-NH-CH-CH-CH_3$ with HOOC and OH	$H-$	6-(Threonylcarbamoyl)purine	—

[a] With permission from Varner, J. E. and D. T. Ho. 1976. Hormones. In: Bonner, J. and J. E. Varner, eds. *Plant Biochemistry.* 3rd ed., pp. 713–770. Copyright by Academic Press, Inc., New York.

Effects of Cytokinins and Other Hormones
in Organisms Other Than Seed Plants

Not only do cytokinins occur in microbial and animal cells, but some interesting effects of exogenous cytokinins on some of these organisms have been reported. For example, cytokinins have been reported to promote the growth (i.e., increase the rate of cell division) of some bacteria, fungi, protozoans, and even cultured mammalian cells. However, it is not clear whether the reported effects of cytokinins in animal cells are directly related to the hormonal effects of cytokinins in plants.

The reported effects of cytokinins in some animal systems point up a fact of more general importance about plant hormones. The mere presence of a substance in an organism does not mean that the substance plays any essential role in the metabolism of that organism. This principle most certainly applies to the substances which we regard as hormones of higher plants. All the major types of hormones, except perhaps abscisic acid, occur in plants of all phyletic groups, but the substances evidently do not have hormone-like activity in all plants. Indeed the "hormone concept" is not even applicable to primitive organisms such as the bacteria. The indoleacetic acid produced in comparative abundance by many bacteria and fungi is not known to be essential to the metabolism of those organisms. GAs produced in relatively large amounts by *Fusarium moniliforme* apparently are unnecessary for the growth of the fungus, although they are important to the host–parasite relationship and pathogenesis into which the fungus enters together with an angiosperm host. There are uncertainties still as to whether auxins and GAs are even essential to the growth of all green plants, such as the algae, for example.

In general, it appears that the hormones of higher plants evolved quite early in comparatively less complex life forms in which they do not have a role directly essential to metabolism—although they are essential to certain types of symbiotic relationships with plant organisms which are phyletically more advanced. The major symptom of the Bakanae disease of rice caused by the fungus *Fusarium moniliforme,* of course, is abnormal elongation of stems and leaf sheaths evoked by the GAs which the pathogen produces and secretes into the host tissue. Fasciation disease, a common disease or pathological overgrowth which attacks a variety of dicotyledonous species, is caused by the cytokinins liberated by the pathogenic bacterium *Corynebacterium tumefaciens* into the host tissues. Extreme cases of this type of pathological condition are called witches brooms, which attack many species and which typically are the result of dozens of lateral buds growing very close together. There are many other cases of symbiotic and host-pathogen relationships between microorganisms and higher plants in which the microbial symbiont or pathogen either liberates hormone into the higher plant partner, or alters the hormone metabolism of the higher plant, or both, causing an abnormal growth response in the higher plant. Examples are the formation of Crown Gall tumors, caused by *Agrobacterium tumefaciens,* and nodulation of legume roots by *Rhizobium* species. The formation of certain galls, caused by particular species of insects, are still other cases in which the hormone metabolism of the higher plants is affected by an invading organism.

Structure/Activity Relationships of the Cytokinins

Previously the cytokinins were characterized as derivatives of adenine, a purine, having a substituent attached to the N^6 position of the adenine ring. Early work indicated that high cytokinin activity was evidently restricted to these N^6-substituted adenines. But an important question which was asked soon after the initial isolation of kinetin was: Just how specific, in fact, are the chemical structural requirements for cytokinin-like biological activity? This is an important question about any type of

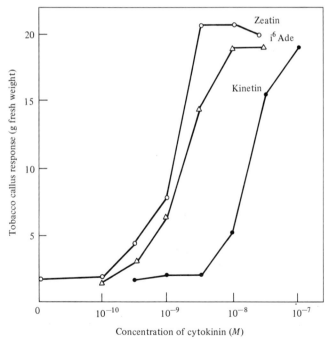

Figure 4-4. Comparative effectiveness of three cytokinins in stimulating growth of tobacco callus in vitro. The compound abbreviated i^6Ade is N^6-(Δ^2-isopentenyl)adenine. (Redrawn with modifications, by permission, from Leonard et al., 1968.)

hormone for at least two reasons: (1) knowing the essential structural features is critical to ultimate elucidation of the specific kind of molecular interaction it undergoes as it exerts hormonal action; and (2) knowing the essential structural features is necessary to efforts to synthesize biologically active analogs and competitive inhibitors of the hormone.

In an extensive study, Skoog et al. (1967) investigated structure/activity relationships of several dozen (69) cytokinin-related substances, most of which were purely synthetic. They found that only N^6-substituted adenine compounds definitely possess cytokinin activity (Fig. 4-4); however, the activity of these compounds varies considerably with the length, degree of unsaturation, and substitutions on the side chain. Modification of the adenine moiety, for example substitution of N for C at the 8-position, reduces activity as measured in the tobacco callus assay by as much as 95%. Even minor substitutions on the adenine nucleus usually reduce activity. A second substituent at the 1- or 3-position of the purine ring nearly eliminates cytokinin activity. However, a second substituent at the N^6, N^7, or N^9 position has much less of a depressing effect. N^9 substituents are metabolically labile, and metabolic conversions of both 3- and 9-substituted derivatives to their corresponding N^6 isomers have been demonstrated. N^6-substituted adenosines possess high cytokinin activity; in some cases they are only slightly less active than their corresponding

Figure 4-5. Three steps that convert the active cytokinin 6-(3-methyl-2-butenyl-amino)purine (2iP or i⁶Ade) (lower right) to a cytokinin antagonist, 3-methyl-7-(3-methylbutylamino)pyrazolo[4,3-d]pyrimidine (upper left). Also growth of tobacco callus on medium containing serial concentrations of each compound. (Reproduced, with permission, from Skoog et al., 1973; photo courtesy of F. Skoog, 1978.)

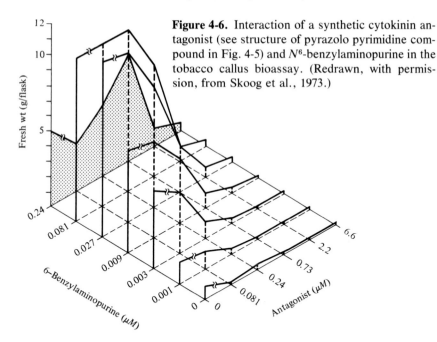

Figure 4-6. Interaction of a synthetic cytokinin antagonist (see structure of pyrazolo pyrimidine compound in Fig. 4-5) and N^6-benzylaminopurine in the tobacco callus bioassay. (Redrawn, with permission, from Skoog et al., 1973.)

adenines. It remains to be fully ascertained, however, whether ribosides and ribotides are active per se or whether cytokinin activity is acquired only upon the conversion to free bases.

Making use of the information on structure/activity relationships, Skoog and associates (1973) synthesized a compound, 3-methyl-7-(3-methylbutylamino)pyrazolo-[4,3-d]pyrimidine (Fig. 4-5), which acts as an active cytokinin antagonist, or a competitive inhibitor of N^6-adenine cytokinins (Fig. 4-6). In Fig. 4-5 three steps are shown which were found to be necessary to convert the cytokinin 6-(3-methyl-2-butenylamino)purine (2iP or i^6Ade) to the pyrazolo pyrimidine antagonist, namely: (1) saturation of the side chain; (2) exchange of the C-9 and N-9 atoms; and (3) addition of a methyl group onto the pyrimidine nucleus. Since the development of the pyrazolo pyrimidine cytokinin antagonist, several other compounds with similar action have been synthesized.

There have been reports that certain nonpurine compounds have cytokinin or cytokinin-like activity, and some workers have suggested that they be considered true cytokinins, although not all physiologists agree. Four such compounds are illustrated in Fig. 4-7. Of these, only N,N'-diphenylurea is naturally occurring; the other three are purely synthetic. Diphenylurea occurs in coconut milk, and at one time the cell division-promoting activity of coconut milk was attributed to that compound. However, as previously noted, Miller (1968) showed that most of the cytokinin activity of coconut milk is attributable to zeatin and its ribonucleoside. Perhaps the so-called "urea cytokinins" are not active as such but

8–Azakinetin

Benzimidazole

N,N′–Diphenylurea

2–Benzthiazolyloxyacetic acid

Figure 4-7. Examples of nonpurine compounds with cytokinin-like activity. [Copyright © 1969 McGraw-Hill Book Co. (UK) Ltd. From The cytokinins by J. Eugene Fox in M. B. Wilkins: *Physiology of Plant Growth and Development.* Reproduced by permission.]

instead serve as inducers of or precursors—i.e., side chain donors—to the N^6-substituted adenine-type cytokinins. Doubt has been expressed about whether 2-benzthiazolyloxyacetic acid really has cytokinin activity; instead it may have auxin activity. 8-Azakinetin and benzimidazole may in fact have cytokinin activity, although the former is only less than 10% as active as kinetin. At the present state of our knowledge, it still seems a valid generalization that the cytokinins are N^6-substituted adenine types of compounds and close analogs.

Biosynthesis and Metabolism

Cytokinin biosynthesis in seed plants evidently occurs generally in tissues and loci which are either meristematic or still retain growth potential. Of particular interest is the fact that cytokinins apparently are synthesized in roots and translocated acropetally in shoots, since cytokinins very frequently have been detected in bleeding sap (xylem sap) of such plants as sunflowers, grapevines, and tobacco plants. This acropetal transport of cytokinins in the vascular tissue appears to be involved in certain so-called "correlative" growth phenomena, such as apical dominance. For example, cytokinin moving acropetally past buds at the nodes evidently opposes the effect of the basipetally moving endogenous auxin. Whereas the auxin inhibits differentiation of the vascular tissues in the bud traces connecting bud and stele, cytokinin promotes differentiation of the bud traces.

Our understanding of the biochemistry of cytokinin production is quite incomplete, and that information which is available is comparatively very complicated. This complexity derives chiefly from the fact that cytokinins occur not only as free bases and the ribonucleosides and ribonucleotides but also as constituents of particular molecular species of tRNA. Most investigations of cytokinin biogenesis in fact have been concerned with the mechanism by which cytokinins come to be present in particular tRNA fractions, and it is this topic to which we will give major consideration.

So far as biosynthesis of the free bases, the cytokinins evidently originate by a condensation of adenine with an appropriate donor of the N^6-substituent. A cell-free enzyme system from the slime mold *Dictyostelium* has recently been reported to form a cytokinin from adenosine monophosphate and Δ^2-isopentenyl pyrophosphate as substrates. The general isoprenoid nature of the N^6-substituents of natural cytokinins indicates that the biosynthesis of the cytokinins parallels, to some extent, the biosynthesis of GAs and abscisic acid.

The metabolic reactions which have been described for the cytokinins include: (1) interconversions, (2) formation of ribosides and ribotides from free bases, (3) glucosylation, (4) methylthiolation, and (5) cleavage of the side chain and breakdown of the purine ring. Examples of interconversions are the formation of N^6-(3-methyl-3-hydroxybutyl)aminopurine from N^6-(3-methyl-2-butenyl)aminopurine (i^6Ade or IPA) and the conversion of the IPA (or i^6Ade) riboside to the zeatin riboside. Formation of the ribosides and ribotides has been demonstrated for zeatin and dihydrozeatin and for the synthetic cytokinins kinetin and benzyladenine. Zeatin also forms glucosides, 7-glucosylzeatin (also called "raphanatin," from *Raphanus sativus*, radish), and 9-glucosylzeatin (see numbering system in Fig. 4-1). In 9-glucosylzeatin the glucose moiety is attached to the 9-position of the purine ring, at which ribose is attached in zeatin nucleoside. In other cases, cytokinin glucosides are formed by the addition of a glucose moiety to the hydroxyl group on the side chain of zeatin. Presently it is not known whether cytokinin ribosides, ribotides, and glucosides are biologically active per se or whether, as seems more likely, they are active only upon liberation of the free bases. If the latter, then some conjugated derivatives—e.g., glucosides—may be important storage forms of cytokinins. The glucosides probably are important also as transport forms in the vascular system. Another type of reaction that cytokinins can undergo is addition of a methylthio-group (CH_3S-) at position 2 of the purine ring (Fig. 4-3). Finally, the N^6-substituent (side chain) may be cleaved, and a residual purine ring may be successively oxidized by xanthine oxidase to yield uric acid and then urea, or carbon 8 may be oxidized away leaving a triaminopurine. Undoubtedly, some or all of these types of metabolic reactions, together with turnover of tRNA containing cytokinin moieties and regulation of biosynthesis, are important in the regulation of levels of active cytokinins in tissues.

Mechanisms of Origin in tRNA

Now to a consideration of the more extensive investigations of the mechanism of origin of cytokinins in tRNA. Cytokinins are unique among plant hormones in that adenine compounds of identical structure occur in nucleic acids. That is, specifically, particular cytokinin-active ribonucleosides occur as components of certain molecular species of tRNA. Several laboratories with quite different directions of research came upon this discovery. Actually, as early as 1960, Thimann and Lalorya first proposed that cytokinins might be rapidly incorporated into macromolecules such as RNA on the basis of their findings that cytokinins promoted RNA and protein synthesis and that exogenous cytokinins often have highly localized effects when applied to whole plants and plant organs. Let us now examine briefly what we know about the occurrence of natural cytokinins in tRNA.

H. G. Zachau et al. (1966), during their determination of the base sequences of serine tRNA in yeast, first reported an "odd" base immediately adjacent to the 3' end of the anticodon (Fig. 4-8). In collaboration with K. Biemann et al. (1966), this "odd" base was identified as the natural cytokinin isopentenyladenosine [(IPA) or 6-(γ,γ-dimethylallyl-amino)purine], which is one of the most highly active cytokinins known. At about the same time, R. H. Hall et al. (1966) reported on the isolation of IPA from crude tRNA preparations. As it was later determined, a cytokinin occurs next to the anticodon actually in all four of the tRNA fractions specific for serine (Fig. 4-9). A short time later, tRNA for tyrosine (tRNA^{Tyr}) was shown by Harada et al. (1968) to contain 2-methylthio-N^6-(Δ^2-isopentenyl)adenosine (ms^{2}i^{6}A) adjacent to the 3' end of the anticodon.

As this research progressed, it came to be known that cytokinins occur in tRNA (certain molecular species only) of a large number of—indeed it is safe to assume now all—organisms. Moreover, in all cases where the specific location of a cytokinin moiety has been determined, it invariably is immediately adjacent to the 3' end of the anticodon. And, interestingly, all tRNA species which are known to contain a cytokinin recognize codons with the initial base as uracil (U) (Fig. 4-9), although not all tRNA species recognizing codons beginning with U contain a cytokinin. The cytokinin-containing tRNA species are tRNA^{Ser}, tRNA^{Phe}, tRNA^{Cys}, tRNA^{Trp}, tRNA^{Leu}, and tRNA^{Tyr} which are specific for the amino acids serine, phenylalanine, cysteine, tryptophan, leucine and tyrosine, respec-

Figure 4-8. Structure of serine-tRNA from Brewers yeast showing the position of isopentenyladenosine adjacent to the 3' end of the anticodon. (Redrawn, with permission, from Zachau et al., 1966b.)

Isopentenyladenosine

tively. In tRNA from microbial sources, cytokinin-active ribonucleosides appear to be present in most, if not all, of the tRNA species that recognize codons beginning with U. But in higher organisms the distribution appears to be more restricted. In the cases of wheat germ and etiolated bean (*Phaseolus vulgaris*) seedlings, cytokinin-active ribonucleosides are limited to only tRNA^{Ser} and tRNA^{Leu}.

Before discussing further the localization of natural cytokinins in tRNA, it should be noted that the natural occurrence of cytokinins is not restricted exclusively to tRNA. As previously noted, cytokinins occur also as the free bases, ribonucleosides and ribonucleotides. Unfortunately there is little information on the relative amounts of cytokinin which occur in these forms, compared to being incorporated into tRNA, for any biological material. It is important to bear in mind that cytokinins occur in multiple chemical forms as the mechanism of hormonal action of the cytokinins is contemplated.

1st Letter	2nd Letter				3rd Letter
	U	C	A	G	
U	PHE	SER	TYR	CYS	U
	PHE	SER	TYR	CYS	C
	LEU	SER	C.T.	C.T.	A
	LEU	SER	C.T.	TRY	G
C	LEU	PRO	HIS	ARG	U
	LEU	PRO	HIS	ARG	C
	LEU	PRO	GLN	ARG	A
	LEU	PRO	GLN	ARG	G
A	ILEU	THR	ASN	SER	U
	ILEU	THR	ASN	SER	C
	ILEU	THR	LYS	ARG	A
	MET (C.I.)	THR	LYS	ARG	G
G	VAL	ALA	ASP	GLY	U
	VAL	ALA	ASP	GLY	C
	VAL	ALA	GLU	GLY	A
	VAL (C.I.)	ALA	GLU	GLY	G

Figure 4-9. Distribution of cytokinin-containing tRNA species in *E. coli*. Striped codons correspond to cytokinin containing tRNA species. Active constituent identified in most cases as 6-(3-methyl-2-butenylamino)-9-β-D-ribofuranosylpurine or derivative. Boxed-in codons correspond to tRNA species reported to contain threoninecarbamoylpurine derivatives. (Reproduced, with permission, from the *Annual Review of Plant Physiology*, Volume 21. © 1970 by Annual Reviews Inc. From Skoog and Armstrong, 1970.)

Acknowledging that natural cytokinins do occur in particular tRNAs, several important questions immediately arise: (1) Are the cytokinins precursors in the formation of tRNA (i.e., is the whole cytokinin molecule incorporated into a molecule of tRNA as the polynucleotide chain is synthesized), or does the presence of a cytokinin moiety in a molecule of tRNA result from the addition of the N^6-substituent to an adenosine residue of preformed tRNA? (2) What, if any, are the metabolic consequences of the presence of cytokinin-active bases in tRNA? (3) Is the observed presence of cytokinin-active bases in tRNA related to the hormonal activity of the free cytokinins? (4) What is the molecular mechanism or mechanisms of action of the cytokinins? Each of these questions will be addressed, but as we do so it will be well to bear in mind that the presence of cytokinins in tRNA does have consequences for the metabolic activity of those tRNAs in protein synthesis, and that, even so, tissues which contain potent cytokinins in their tRNA (e.g., tobacco callus) still require exogenous cytokinin for growth in vitro. Should these facts necessarily lead us to speculate that free cytokinins have a biological role or mechanism of action quite distinct from the cytokinin moieties which occur in tRNA? Later discussion will aid in trying to answer this as well as the preceding questions.

Now, as to the first question posed above. Are the cytokinins incorporated intact into tRNA, in a manner like all other nucleoside triphosphates that are polymerized in the synthesis of tRNA, or does the presence of the cytokinins result from alkylation (transfer of isoprenoid side chain) of existing adenosine moieties already present in preformed tRNA? In fact, experimental evidence has been reported for both processes. In 1966 Eugene Fox of the University of Kansas reported that ^{14}C label from the synthetic cytokinin N^6-benzyladenine (whether labeled in the 8-position of the adenine moiety or the methylene carbon of the benzyl side chain) was incorporated into RNA in soybean and tobacco tissue cultures, both of which have an absolute cytokinin requirement. Further work by Fox and associates and workers in Skoog's group at Wisconsin has extended that early report considerably. Evidence was presented that two synthetic cytokinins—kinetin (N^6-furfuryladenine) and benzyladenine—are incorporated into RNA of cytokinin-dependent tobacco callus tissue. Moreover, it was shown that N^6-benzyladenine is incorporated as the intact molecule into both tRNA and ribosomal RNA (rRNA). Indeed, both synthetic cytokinins were recovered as the corresponding ribonucleosides to about three times greater extent per absorbance unit at 260 nm from the rRNA than from the tRNA preparations. But the amount of incorporation was very small, and it remains uncertain yet whether the observed incorporation of exogenous cytokinins into RNA is the result of specific events involved in the mechanism of cytokinin action or merely the result of transcriptional errors.

Meanwhile, the evidence for cytokinin moieties in tRNA arising by alkylation of specific adenosine residues in preformed tRNA is conclusive. For example, in the bacterium *Escherichia coli* N^6-(Δ^2-isopentenyl)adenosine (i^6A) and 6-(3-methyl-2-butenylamino)-2-methylthio-9-β-D-ribofuranosylpurine (ms^2i^6A) have been identified as cytokinin-active ribonucleosides in tRNA, and both cytokinin moieties are located next to the 3' end of the anticodon of tRNA species recognizing codons beginning with uridine. In this system, it is known definitely that the cytokinin i^6A is synthesized by the transfer of the isopentenyl moiety from Δ^2-isopentenyl pyrophosphate to the specific adenosine residues adjacent to the 3' end of the anticodon in preformed tRNA molecules. Moreover, as will be discussed later, there is good evidence from the *E. coli* system that the cytokinin-active components of tRNA function to ensure proper codon–anticodon interaction of the mRNA–aminoacyl tRNA complex on ribosomes. Also in tobaccco callus—indeed for all organisms examined thus far—this process of transfer of an isopentenyl group to specific adenosine residues in preformed tRNA molecules is a conclusively demonstrated means by which cytokinin-active moieties come to be present in specific tRNA molecules. At this stage of our knowledge, therefore, we reasonably can conclude that the latter process accounts for the presence of cytokinins in certain tRNAs, but we cannot say whether it is the only way.

Metabolic Consequences of the Presence of Cytokinins in tRNA

Regarding another question posed previously, it does indeed seem that there are significant metabolic consequences to the occurrence of cyto-kinins in tRNA. That is to say, there is good evidence that the cytokinin moieties in tRNA are functionally significant and that they do affect the behavior of those tRNA molecules in the process of protein synthesis. One of the earliest and most conclusive investigations from which this important fact emerged was conducted by M. L. Gefter and R. L. Russell (1969) with *Escherichia coli*.

Gefter and Russell (1969) infected *E. coli* cells with a "defective trans-ducing bacteriophage" carrying the tyrosine tRNA gene. This caused an amplification of the production by *E. coli* of actually three molecular spe-cies of tRNA for tyrosine (tRNA^{Tyr}). The three forms of tRNA^{Tyr} had the same nucleotide sequence and differed only in the degree of modification of the adenosine residue adjacent to the 3' end of the anticodon: (1) un-modified adenosine (A), (2) N^6-(Δ^2-isopentenyl)adenosine (i^{6}A), and (3) 6-(3-methyl-2-butenylamino)-2-methylthio-9-β-D-ribofuranosylpurine (ms^{2}i^{6}A). All three forms of the tRNA were then tested for tyrosine accep-tor activity and for binding of the tyrosyl-tRNA to an mRNA–ribosome complex. The results were very interesting. No significant differences were found among the three forms of tRNA^{Tyr} as regards amino acid ac-ceptor capacity. However, the tRNA^{Tyr} containing an unmodified adeno-sine residue adjacent to the anticodon was markedly less effective, in the in vitro experiments, in the binding of the amino acyl–tRNA^{Tyr} to the mRNA–ribosome complex than tRNA^{Tyr} containing i^{6}A or ms^{2}i^{6}A. Thus, in this very important case, a cytokinin moiety evidently must be present adjacent to the anticodon of tRNA^{Tyr} if this tRNA species is to function effectively in protein synthesis.

Somewhat earlier, Fittler and Hall (1966) also found that a chemical change in a cytokinin moiety located in tRNA modified the activity of that tRNA in protein synthesis. Specifically, they observed that iodination (treatment with aqueous iodine) of the tRNA for serine of yeast decreased the binding affinity of the seryl-tRNA for the mRNA–ribosome complex without affecting the acceptor activity for serine.

In contrast, other investigators produced both normal and i^{6}A-deficient tRNAs from *Lactobacillus acidophilus* by limiting the mevalonate uptake in the latter to that amount which was just sufficient for maximum growth of the bacteria. Under these conditions, the i^6A content of the tRNA was reduced to about half the normal amount. Yet both the normal and under-modified tRNAs were about equally effective in amino acid acceptor activity and protein synthesis in vitro.

In summary, clearly the cytokinin moieties in particular species of tRNA are functionally significant. It appears that the cytokinin moieties have an important regulatory function for tRNA. It is not known by what mechanism that regulation is achieved. Perhaps it is by preventing a wrong set of three nucleotides from being recognized by a codon, but this is only one of several speculative ideas. By whatever mechanism the regulation is achieved, however, it appears that cytokinin-active bases in certain types of tRNA may have a modulating effect on protein synthesis at the translational level. However, the important question remains as to whether the reported action of cytokinin moieties in tRNA is related, directly or indirectly, to the hormonal action of free cytokinins.

Hormonal Activity of Free Cytokinins

There are good reasons presently for believing that free cytokinins have important biological activity independently of any direct association with tRNA: (1) there are some investigations, to be described shortly, which have been interpreted as showing directly biological and biochemical activity of exogenous cytokinin independently of incorporation into tRNA; (2) ethanolic extracts of corn (*Zea mays*) kernels contain the *trans*-isomer of zeatin, while the tRNA hydrolysates of corn kernels contain the *cis*-isomer, suggesting that zeatin is not a precursor in the synthesis of tRNA in that material; (3) likewise, dihydrozeatin is a major free cytokinin in beans (*Phaseolus vulgaris*), but it apparently does not occur in bean tRNA; (4) even tissues which contain potent cytokinins in their RNA still require exogenous cytokinin for growth in vitro; and (5) the data on direct incorporation of exogenous cytokinins into tRNA do not show conclusively that this process is anything more than transcriptional error.

A prevalent idea about mechanisms of action of plant hormones, cytokinin free bases and others, is that the hormone first binds—by weak hydrogen or ionic bonds—to some receptor. The receptor commonly is envisaged to be an allosteric protein which, as a consequence of binding the hormone, can, as the altered receptor or as a receptor–hormone complex, evoke hormonal action. This idea, well established in animal physiology, has found experimental support in plant physiology only comparatively quite recently (see review by Kende and Gardner, 1976).

Since about 1970, some progress along these lines with respect to cytokinins has been made by J. Eugene Fox and associates at the University of Kansas and others. A basic observation that Fox and Erion (1975) made was that cytokinins would bind, with rather high specificity, to ribosomes isolated from higher plants. Extending that observation, Fox and Erion (1977) described actually three cytokinin-binding proteins isolated from ribosomal preparations from wheat germ. They found a fraction of medium molecular weight of 93,000 (CBF-1) with high binding affinity, a low-molecular weight fraction (30,000) with a high affinity for cytokinins (CBF-2), and a large-molecular weight fraction of more than 250,000 which had a low affinity for cytokinins (CBF-3). The CBF-1 fraction was isolated specifically from ribosomes, although it was found to occur also in the high-speed postribosomal supernatant together with CBF-2 and CBF-3. A large number of cytokinins, cytokinin nucleotides, analogs, and related purines, and other plant growth substances were tested for their ability to compete with 6-benzylaminopurine (or N^6-benzyladenine) for binding sites on CBF-1. Apparently wheat germ ribosomes contain one copy of CBF-1 per ribosome. The competition studies showed a high degree of specificity for cytokinin-active moieties. The biological role of these cytokinin-binding proteins presently is unknown, but it will be very interesting to follow further research in this area. The binding of cytokinins to ribosomes conceivably could have profound consequences for the regulation of protein synthesis.

Effects on Moss Protonemata

Some of the most intriguing data on biological effects of exogenous cytokinins have come from investigations of "bud" (actually gametophore) production by protonemata of mosses (Figs. 4-10, 4-11, and 4-12). Evidently "bud" formation in these cryptogams is naturally regulated by an endogenous adenine-type of cytokinin. The haploid moss spore germinates to form a green algal-like filament called a chloronema, which develops into a caulonema, and it is from subapical cells of the caulonema that the "buds" which give rise to gametophores are formed. The chloronema and caulonema are but sequential stages in the development of the green, branching, septate protonema.

H. Brandes and Hans Kende of the MSU/DOE Plant Research Laboratory expanded and quantified some of the early observations obtained with the moss system (Figs. 4-10, 4-11, 4-12). Observations were made which were interpreted as showing that the synthetic cytokinin when supplied exogenously to protonemata of *Funaria hygrometrica* was active independently of any covalent bonding of the hormone. Using ^{14}C-labeled N^6-benzyladenine, they found that during early stages of their development moss "buds" revert to protonemal filaments if protonemata were washed in an effort to remove the free cytokinin. These results were interpreted further as indicating that the exogenous cytokinin was not acting as

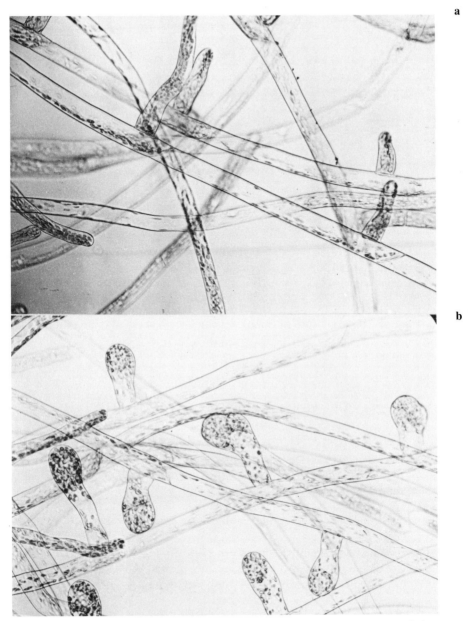

Figure 4-10. Effects of benzyladenine on bud formation in protonemata of the moss *Funaria hygrometrica*. **a**, Filaments of a protonema grown on basal medium for 15 days. **b**, Bud formation on filaments of a protonema which was cultured on basal medium for 14 days and then transferred to basal medium containing 1 μM benzyladenine for 24 hours. (With permission, from Brandes and Kende, 1968; photos courtesy of H. Kende, 1978.)

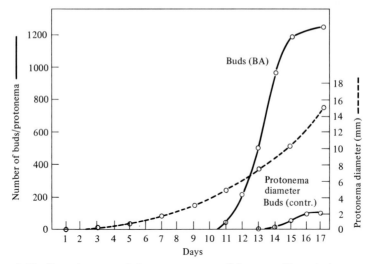

Figure 4-11. Growth curve of the protonemata of the moss *Funaria hygrometrica* in vitro and spontaneous and benzyladenine-induced (1 μM) bud formation. (Redrawn, with permission, from Brandes and Kende, 1968.)

a "trigger" but that, like hormone responses generally, the cytokinin had to be present during a critical period of time until "bud" differentiation was stabilized.

Some Physiological Effects on Seed Plants

Like auxins, gibberellins, and all other plant hormones, the cytokinins have many diverse physiological effects on seed plants, some of which have been mentioned previously. This diversity of physiological effects complicates efforts to elucidate molecular mechanisms of hormones. However, such diversity of effects is only natural and logically to be expected when it is realized that, in fact, the regulation of growth and development in plants is achieved by several types of hormones acting simultaneously. Hence it really is not surprising to learn that cytokinins have been demonstrated to be involved in regulation of seed plants literally throughout their ontogeny. Of these multiple effects only three of the major ones will be discussed in this section.

Effects of Auxin and Cytokinin on Morphogenesis in Tobacco Callus Tissue

In a classic paper published in 1957, Skoog and Miller described a unique example of hormonal regulation of morphogenesis in tobacco pith explants (Fig. 4-13). This important research beautifully documented a delicate and quantitative interaction between auxin and cytokinin in the con-

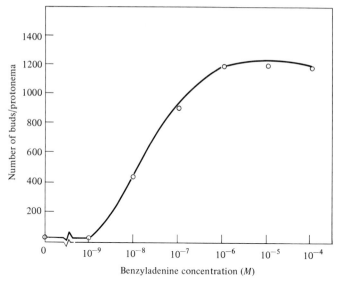

Figure 4-12. Dose–response curve for benzyladenine effect on bud formation in moss protonemata. (Redrawn, with permission, from Brandes and Kende, 1968.)

trol of bud and root formation from pith explants. They showed that with a particular combination of concentrations of IAA and kinetin, the pith tissues grew as relatively undifferentiated callus. By varying the ratio of

Figure 4-13. Effects of auxin (IAA) and cytokinin (kinetin) on growth and morphogenesis in tobacco callus. (With permission, from Skoog and Miller, 1957; photo courtesy of F. Skoog, 1978.)

IAA:kinetin, however, they could successfully cause the explants to give rise to buds or to roots. Thus by varying the amounts of the two types of growth substances in the culture medium, morphogenesis in tobacco pith explants can be controlled to a remarkable degree. The implications for this kind of research for understanding normal, as well as abnormal or pathological, growth are great indeed.

Effects of Cytokinins on Leaf Growth and Senescence

Senescence is one of the most poorly defined and least understood—yet very important—phenomena in biology. By a generally acceptable connotation, senescence refers to the collective progressive and deteriorative processes which naturally terminate in the death of an organ or organism. In this sense "senescence" is synonymous with "aging."

There are many familiar manifestations of senescence in higher plants. Whole plant senescence is obvious in annual plants in which the life cycle is terminated at the end of one growing season. In herbaceous perennials, the aerial portion of the shoot senesces and dies at the end of a growing season, but the underground system remains viable. Annual senescence and abscission of leaves is exhibited by many woody perennials. Progressive senescence and death of leaves from the basal portion acropetally on a shoot is common. Finally, the ripening and subsequent deterioration of fleshy fruits is an example of senescence.

The biochemical changes that occur in leaves as they grow old have been characterized extensively. However, the endogenous factors which regulate these changes and the reasons why the leaf eventually dies are incompletely understood. The most prominent changes that occur in the senescent leaf are declines in the proteins and nucleic acids and an irreversible yellowing due to the loss of chlorophyll. Catabolism exceeds anabolism, and there is massive export of a variety of soluble metabolites out of the senescent leaf to other parts of the plant. Photosynthesis declines, and generally the respiration rate decreases also. Cytological changes include structural, as well as functional, disorganization of organelles, increased membrane permeability, destruction of cytoplasmic polyribosomes, and eventual disintegration of nuclei.

Leaf senescence and abscission (in those species in which abscission occurs) actually is a conservative biological phenomenon. Lower (older) leaves on a shoot axis often become quite inefficient organs of photosynthesis because of the shading by the upper portion of the shoot. Thus death and possible elimination of the organ, preceded by export of soluble metabolites, is a conservative process which contributes to the overall welfare of the plant.

When a mature leaf is excised from the plant and the petiole (or basal part) is submerged basally in water, senescence occurs rapidly, provided the petiole does not form adventitious roots. If adventitious roots form on the petiole of an excised leaf rapid senescence, typical of the excised mature leaf, does not occur, because the roots export endogenous cytokinin to the leaf in the xylem sap. The protein and chlorophyll content of the nonrooted leaf decline to less than half the original value in a matter of just a few days. The same process can be observed with excised sections or disks of leaves floated on water.

In 1957, A. E. Richmond and A. Lang discovered that kinetin retarded the senescence of detached leaves of cocklebur (*Xanthium*). This discovery was followed rapidly by numerous reports showing that kinetin and benzyladenine, another synthetic cytokinin, retard senescence of whole excised leaves and excised sections of leaves of numerous species. If whole excised leaves are quite young when excised, cytokinin stimulates growth appreciably.

This area of research was advanced dramatically in 1959 and the years following by K. Mothes and associates in Germany. They found by spraying solutions of kinetin directly onto leaves that the effect of the applied cytokinin was quite localized. Only those areas to which the chemical was applied remained green. Furthermore, the treated areas of yellowing leaves actually became greener. In an effort to explain these intriguing observations, Mothes and associates applied radioactive amino acids and nonradioactive kinetin to leaves. They found that radioactive amino acids migrated to and accumulated in the kinetin-treated parts of the leaves. Thus it came to be understood that cytokinin causes mobilization of metabolites from an untreated to a treated portion of a leaf—i.e., creates new source-sink relationships—and that the chemical somehow acts to sustain nucleic acid and protein synthesis.

A. C. Leopold and M. Kawase (1964) advanced this research a step further using shoot cuttings of bean (*Phaseolus vulgaris*) seedlings. They observed another interesting effect of benzyladenine in addition to stimulation of growth of intact whole young leaves and retardation of senescence of localized treated parts of mature leaves. Treatment of one leaf with cytokinin actually caused inhibition of growth and hastened senescence of other, untreated leaves on the cuttings (Fig. 4-14).

Roles of Cytokinins and Auxin in Apical Dominance

"Apical dominance" refers to the inhibition of growth of subtending lateral (axillary) buds by a growing shoot apex. Some degree of apical dominance is found in all seed plants, and the phenomenon obviously is of profound significance in determining growth form. In some herbaceous plants [e.g., pea (*Pisum sativum*) and sunflower (*Helianthus annuus*)] apical dominance is quite strong, the lateral buds commonly are suppressed, and the shoot thus develops as a single main axis. In other species [e.g., tomato (*Lycopersicum esculentum*) and potato (*Solanum tuberosum*)] apical dominance is much weaker and the shoot is extensively branched. Sometimes apical dominance is only partial, in which case some of the

Figure 4-14. Senescence of a trifoliate leaf of a beam (*Phaseolus vulgaris*) shoot cutting caused by treating the subtending pair of primary leaves with benzyladenine. (With permission, from Leopold and Kawase, 1964; photo courtesy of A. C. Leopold, 1978.)

lateral buds develop while others remain fully inhibited.

In certain species of vascular plants—e.g., *Ginkgo biloba,* the Maidenhair tree, and *Cercidiphyllum japonicum,* the Katsura tree of Japan—there is a special and very interesting case of lateral buds which open but which do not elongate appreciably. Instead they produce "short shoots" as opposed to the more usual "long shoots." When a vigorous terminal bud develops first it establishes dominance over the laterals below. This inhibition is transmitted by the auxin produced in the elongating terminal ("long") shoot. In these plants a lateral "short shoot" behaves fundamentally like an inhibited bud of an herbaceous plant, differing in that it opens while an inhibited lateral bud does not. Therefore the occurrence of "short shoots" and "long shoots" on a plant such as *Ginkgo* is regarded as a special instance of the phenomenon of apical dominance.

The two cases just described indicate that apical dominance is very important in determining the growth habits of woody perennials. However, some confusion has existed in the past because of attempts to explain the adult forms of trees and shrubs by too strict and sometimes erroneous analogy with apical dominance in herbaceous plants. The relationship between bud inhibition and form in woody plants is much more complex than bud inhibition in herbaceous plants because of the time sequence in the formation and release of lateral buds. The life cycles of woody perennials are characterized by periods of dormancy. A dormant apical bud does not exert an inhibitory influence on the subtending lateral buds, and more than one bud may develop when dormancy is broken and growth resumes. This may partially explain why so much of the branching of woody perennials occurs at the beginning of a period of active growth. Apical dominance most certainly is directly involved in determining the pattern of growth of individual twigs and branches. However, it probably would be well, as suggested by C. L. Brown et al. (1967), to restrict the term to the pattern of bud inhibition in currently elongating twigs, and to denote the complex of physiological conditions giving rise to excurrent and decurrent growth forms of whole trees and shrubs, as "apical control."

While the mechanism of apical dominance is incompletely understood at present, K. V. Thimann and T. Sachs have showed that the phenomenon involves an antagonism between cytokinins and auxins. Auxin, produced at the shoot tip and transported basipetally, is inhibitory to the release of buds from apical dominance, and this action of auxin is antagonized by cytokinin, some of which is synthesized in the suppressed lateral bud itself, and some of which is transported past the bud in the xylem sap. Lateral buds typically are relieved of inhibition in a growing shoot experimentally by removal of the shoot apex. Inhibition of the lateral buds is reinstated if auxin is applied to the cut surface of the decapitated stem. Local application of cytokinin directly to inhibited lateral buds causes a release from inhibition. However, buds thus released from apical dominance do not elongate as much as uninhibited buds released by decapitation of the shoot apex, unless they are treated with auxin following release with cytokinin. Thus, in the intact plant, the inhibited lateral bud may—as paradoxical as it might at first appear—fail to grow because of a deficiency in both cytokinin and auxin.

Correlated directly with the effects of auxin and cytokinin on lateral bud inhibition and release therefrom are well-documented effects of these two kinds of hormones on differentiation of vascular tissues between axillary buds and the primary stele of the main shoot axis. Histological studies have revealed that inhibited lateral buds of some plants lack well-developed vascular connections with the vascular system of the main stem. This observation suggests, of course, that inhibited buds receive a deficient supply of metabolites, which evidence shows actually to be the case. Removal of the apical bud from a growing shoot leads to rapid development of vascular connections between the released bud and the vascular system of the main shoot axis. As would be predicted, auxin applied to a decapitated shoot tip inhibits the formation of vascular connections, and cytokinin applied locally to the lateral buds promotes the differentiation of the vascular traces.

Translocation

Somewhat of a paradox exists with regard to translocation of the cytokinins. On the one hand, there is abundant data which reveal that exogenous cytokinins, when applied to leaves, stems, and buds, exhibit markedly little if any movement from the site of application. This relative immobility is an essential feature of the so-called "mobilization effect" of cytokinins, according to which a localized application of exogenous cyto-kinin to a leaf or part of a leaf delays senescence in that localized portion and creates a sink to which metabolites are transported from other parts of the leaf or even other leaves (see Fig. 4-14).

On the other hand, however, it is well documented that xylem sap and phloem sap as well contain cytokinins. The ability of cytokinins to move upward in the xylem sap has been confirmed by detecting cytokinins in the exudate from decapitated root systems, and by injecting exogenous cytokinins into stems and measuring the resulting systemic effects. Cyto-kinins, especially as glucosides, also occur naturally in phloem sap, as in-dicated, for example, by their occurrence in aphid honeydew.

Possibly these apparently conflicting observations really are not irrec-oncilable. Cytokinins present in the vascular tissues move passively with other solutes in strict accordance with source-to-sink relationships. The major sinks for cytokinins are parts of the plant which are meristematic or which otherwise have growth potential, such as young leaves, buds, young internodes, developing seeds, and fruits—parts which also are the primary centers of cytokinin production together, evidently, with root tips. Moreover, when physiologists apply exogenous cytokinins to intact whole plants or plant organs, they are most often applied to the same sink areas or areas of cytokinin synthesis, because those are the parts which are most responsive.

References

Alvim, R., E. W. Hewett, and P. F. Saunders. 1976. Seasonal variation in the hor-mone content of willow. I. Changes in abscisic acid content and cytokinin ac-tivity in the xylem sap. *Plant Physiol.* **57**: 474–476.

Armstrong, D. J., N. Murai, B. J. Taller, and F. Skoog. 1976. Incorporation of the cytokinin N^6-benzyladenine into tobacco callus transfer ribonucleic acid and ribosomal ribonucleic acid preparations. *Plant Physiol.* **57**: 15–22.

Beimann, K., S. Tsunakawa, J. Sonnenbichler, H. Feldmann, D. Dütting, and H. G. Zachau. 1966. Structure of an odd nucleoside from serine-specific trans-fer ribonucleic acid. *Angew. Chem. Int. Edit.* **5**: 590–591.

Black, M. K. and D. J. Osborne. 1965. Polarity of transport of benzyladenine, adenine and indole-3-acetic acid in petiole segments of *Phaseolus vulgaris*. *Plant Physiol.* **40**: 676–680.

Brandes, H. and H. Kende. 1968. Studies on cytokinin-controlled bud formation in moss protonemata. *Plant Physiol.* **43**: 827–837.

Brown, C. L., R. G. McAlpine, and P. P. Kormanik. 1967. Apical dominance and form in woody plants: a reappraisal. *Am. J. Bot.* **54**: 153–162.

Burrows, W. J. 1975. Mechanism of action of cytokinins. *Current Adv. Plant Sci.* **7**: 837–847.

Chen, C.-M. and R. H. Hall. 1969. Biosynthesis of N^6-(Δ^2-isopentenyl)adenosine in the transfer ribonucleic acid of cultured tobacco pith tissue. *Phytochemistry* **8**: 1687–1695.

Crafts, C. B. and C. O. Miller. 1974. Detection and identification of cytokinins produced by mycorrhizal fungi. *Plant Physiol.* **54**: 586–588.

Deleuze, G. G., J. D. McChesney, and J. E. Fox. 1972. Identification of a stable cytokinin metabolite. *Biochem. Biophys. Res. Commun.* **48**: 1426–1432.

Doree, M. and J. Guern. 1973. Short-time metabolism of some exogenous cytokinins in *Acer pseudoplatanus* cells. *Biochim. Biophys. Acta* **304**: 611–622.

Dyson, W. H., J. E. Fox, and J. D. McChesney. 1972. Short term metabolism of urea and purine cytokinins. *Plant Physiol.* **49**: 506–513.

Einset, J. W. and F. Skoog. 1973. Biosynthesis of cytokinins in cytokinin-autotrophic tobacco callus. *Proc. Natl. Acad. Sci. U.S.A.* **70**: 658–660.

Fittler, F. and R. H. Hall. 1966. Selective modification of yeast seryl-tRNA and its effect on the acceptance and binding functions. *Biochem. Biophys. Res. Commun.* **25**: 441–446.

Fittler, F., L. K. Kline, and R. H. Hall. 1968. Biosynthesis of N^6-(Δ^2-isopentenyl)adenosine. The precursor relationship of acetate and mevalonate to the Δ^2-isopentenyl group of the transfer ribonucleic acid of microorganisms. *Biochemistry* **7**: 940–944.

Fox, J. E. 1966. Incorporation of a kinin, N,6-benzyladenine into soluble RNA. *Plant Physiol.* **41**: 75–82.

Fox, J. E. 1969. The cytokinins. In: Wilkins, M. B., ed. *The Physiology of Plant Growth and Development*. McGraw-Hill Book Publishing Company Limited, London. Pp. 85–123.

Fox, J. E. and C.-M. Chen. 1967. Characterization of labeled ribonucleic acid from tissue grown on ^{14}C-containing cytokinins. *J. Biol. Chem.* **242**: 4490–4494.

Fox, J. E., J. Cornette, G. Deleuze, W. Dyson, C. Giersak, P. Niu, J. Zapata, and J. McChesney. 1973. The formation, isolation, and biological activity of a cytokinin 7-glucoside. *Plant Physiol.* **52**: 627–632.

Fox, J. E. and J. L. Erion. 1975. A cytokinin binding protein from higher plant ribosomes. *Biochem. Biophys. Res. Commun.* **64**: 694–700.

Fox, J. E. and J. Erion. 1977. Cytokinin-binding proteins in higher plants. In: Pilet, P. E., ed. *Plant Growth Regulation*. Springer-Verlag, New York. Pp. 139–146.

Fox, J. E., C. K. Sood, B. Buckwalter, and J. D. McChesney. 1971. The metabolism and biological activity of a 9-substituted cytokinin. *Plant Physiol.* **47**: 275–281.

Gefter, M. L. and R. L. Russell. 1969. Role of modifications in tyrosine transfer RNA: A modified base affecting ribosome binding. *J. Mol. Biol.* **39**: 145–157.

Hall, R. H. 1970. N^6-(Δ^2-isopentenyl)adenosine: Chemical reactions, biosynthesis, metabolism, and significance to the structure and function of tRNA. *Prog. Nucl. Acid Res. Mol. Biol.* **10**: 57–86.

Hall, R. H. 1973. Cytokinins as a probe of developmental processes. *Ann. Rev. Plant Physiol.* **24**: 415–444.

Hall, R. H., L. Csonka, H. David, and B. McLennan. 1967. Cytokinins in the soluble RNA of plant tissues. *Science* **156**: 69–71.

Hall, R. H. and G. Mintsioulis. 1973. Enzymatic activity that catalyzes degradation of N^6-(Δ^2-isopentenyl)adenosine. *J. Biochem.* **73**: 739–748.

Hall, R. H., M. J. Robins, L. Stasiuk, and R. Thedford. 1966. Isolation of ^{6}N-(γ,γ-dimethylallyl)adenosine from soluble ribonucleic acid. *J. Am. Chem. Soc.* **88**: 2614–2615.

Harada, F., H. J. Gross, F. Kimura, S. H. Chang, S. Nishimura, and U. L. Raj Bhandary. 1968. 2-Methylthio-N⁶-(Δ²-isopentenyl)adenosine: a component of *E. coli* tyrosine transfer RNA. *Biochem. Biophys. Res. Commun.* **33**: 299–306.

Harvey, B. M. R., B. C. Lu, and R. A. Fletcher. 1974. Benzyladenine accelerates chloroplast differentiation and stimulates photosynthetic enzyme activity in cucumber cotyledons. *Can. J. Bot.* **52**: 2581–2586.

Helgeson, J. P. and N. J. Leonard. 1966. Cytokinins: Identification of compounds isolated from *Corynebacterium fascians. Proc. Natl. Acad. Sci. U.S.A.* **56**: 60–63.

Horgan, R., E. W. Hewett, J. G. Purse, J. M. Horgan, and P. F. Wareing. 1973. Identification of a cytokinin in sycamore sap by gas chromatography-mass spectrometry. *Plant Sci. Lett.* **1**: 321–324.

Itai, C. and Y. Vaadia. 1965. Kinetin-like activity in root exudate of water-stressed sunflower plants. *Physiol. Plantarum* **18**: 941–944.

Jablonski, J. R. and F. Skoog. 1954. Cell enlargement and cell division in excised tobacco pith tissue. *Physiol. Plantarum* **7**: 16–24.

Kende, H. 1964. Preservation of chlorophyll in leaf sections by substances obtained from root exudate. *Science* **145**: 1066–1067.

Kende, H. 1965. Kinetin like factors in the root exudate of sunflowers. *Proc. Natl. Acad. Sci. U.S.A.* **53**: 1302–1307.

Kende, H. and G. Gardner. 1976. Hormone binding in plants. *Ann. Rev. Plant Physiol.* **27**: 267–290.

Klambt, D., G. Thies, and F. Skoog. 1966. Isolation of cytokinins from *Corynebacterium fascians. Proc. Natl. Acad. Sci. U.S.A.* **56**: 52–59.

Lagerstedt, H. B. and R. G. Langston. 1967. Translocation of radioactive kinetin. *Plant Physiol.* **42**: 611–622.

Leonard, N. J. 1974. Chemistry of the cytokinins. In: Runeckles, V. C., E. Sondheimer, and D. C. Walton, eds. *The Chemistry and Biochemistry of Plant Hormones.* Vol. 7. *Recent Advances in Phytochemistry.* Academic Press, New York. Pp. 21–56.

Leonard, N. J., S. M. Hecht, F. Skoog, and R. Y. Schmitz. 1968. Cytokinins: synthesis of 6(3-methyl-3-butenylamino)9-8-D-ribofuranosylpurine (3iPA), and the effect of side-chain unsaturation on the biological activity of isopentylaminopurines and their ribosides. *Proc. Natl. Acad. Sci. U.S.A.* **59**: 15–21.

Leopold, A. C. and M. Kawase. 1964. Benzyladenine effects on bean leaf growth and senescence. *Am. J. Bot.* **51**: 294–298.

Leopold, A. C. and P. E. Kriedemann. 1975. *Plant Growth and Development. 2nd ed.* McGraw-Hill Book Company, New York.

Letham, D. S. 1963. Zeatin, a factor inducing cell division isolated from *Zea mays. Life Sci.* **2**: 569–573.

Letham, D. S. 1967. Chemistry and physiology of kinetin-like compounds. *Ann. Rev. Plant Physiol.* **18**: 349–364.

Letham, D. S. 1973. Transfer RNA and cytokinins. In: Stewart, P. R. and D. S. Letham, eds. *The Ribonucleic Acids.* Springer-Verlag, New York. Pp. 81–106.

Letham, D. S. and C. O. Miller. 1965. Identity of kinetin-like factors from *Zea mays. Plant Cell Physiol.* **6**: 355–359.

Letham, D. S. and R. K. Ralph. 1967. A cytokinin in soluble RNA from a higher plant. *Life Sci.* **6**: 387–394.

Letham, D. S., J. S. Shannon, and I. R. McDonald. 1964. The structure of Zeatin, a factor inducing cell division. *Proc. Chem. Soc.* 230–231.

Matsubara, S., D. J. Armstrong, and F. Skoog. 1968. Cytokinins in tRNA of *Corynebacterium fascians*. *Plant Physiol.* **43**: 451–453.

Miller, C. O. 1961a. Kinetin and related compounds in plant growth. *Ann. Rev. Plant Physiol.* **12**: 395–408.

Miller, C. O. 1961b. A kinetin-like compound in maize. *Proc. Natl. Acad. Sci. U.S.A.* **47**: 170–174.

Miller, C. O. 1974. Ribosyl-*trans*-zeatin, a major cytokinin produced by crown gall tumor tissue. *Proc. Natl. Acad. Sci. U.S.A.* **71**: 334–338.

Miller, C. O., F. Skoog, F. S. Okumura, M. H. von Saltza, and F. M. Strong. 1955. Structure and synthesis of kinetin. *J. Am. Chem. Soc.* **77**: 2662–2663.

Miller, C. O., F. Skoog, M. H. von Saltza, and F. M. Strong. 1955. Kinetin, a cell division factor from deoxyribonucleic acid. *J. Am. Chem. Soc.* **77**: 1392.

Miura, G. A. and C. O. Miller. 1969. 6-($\gamma\gamma$-dimethylallylamino)purine as a precursor of zeatin. *Plant Physiol.* **44**: 372–376.

Mothes, K. and L. Engelbrecht. 1961. Kinetin-induced directed transport of substances in excised leaves in the dark. *Phytochemistry* **1**: 58–62.

Parker, C. W., M. M. Wilson, and D. S. Letham. 1973. The glucosylation of cytokinins. *Biochem. Biophys. Res. Commun.* **55**: 1370–1376.

Person, C., D. J. Samborski, and F. R. Forsyth. 1957. Effect of benzimidazole on detached wheat leaves. *Nature (London)* **180**: 1294–1295.

Peterkofsky, A. 1968. The incorporation of mevalonic acid into the N^6-(Δ^2-isopentenyl)adenosine of transfer ribonucleic acid in *Lactobacillus acidophilus*. *Biochemistry* **7**: 472–482.

Phillips, I. D. J. 1969. Apical dominance. In: Wilkins, M. B., ed. *The Physiology of Plant Growth and Development*. McGraw-Hill Book Company, New York. Pp. 165–202.

Phillips, I. D. J. 1971. *Introduction to the Biochemistry and Physiology of Plant Growth Hormones*. McGraw-Hill Book Company, New York.

Richmond, A. E. and A. Lang. 1957. Effect of kinetin on protein content and survival of detached Xanthium leaves. *Science* **125**: 650–651.

Sachs, T. and K. V. Thimann. 1964. Release of lateral buds from apical dominance. *Nature (London)* **201**: 939–940.

Sachs, T. and K. V. Thimann. 1967. The role of auxins and cytokinins in the release of buds from dominance. *Am. J. Bot.* **54**: 136–144.

Scarbrough, E., D. J. Armstrong, F. Skoog, C. R. Frihart, and N. J. Leonard. 1973. Isolation of *cis*-zeatin from *Corynebacterium fascians* cultures. *Proc. Natl. Acad. Sci. U.S.A.* **70**: 3825–3829.

Shaw, G. and D. V. Wilson. 1964. Synthesis of zeatin. *Proc. Chem. Soc.* 231.

Shibaoka, H. and K. V. Thimann. 1970. Antagonisms between kinetin and amino acids. Experiments on the mode of action of cytokinins. *Plant Physiol.* **46**: 212–220.

Skene, K. G. M. 1975. Cytokinin production by roots as a factor in the control of plant growth. In: Torrey, J. G. and D. T. Clarkson, eds. *The Development and Function of Roots*. Academic Press, New York. Pp. 365–396.

Skoog, F. and D. J. Armstrong. 1970. Cytokinins. *Ann. Rev. Plant Physiol.* **21**: 359–384.

Skoog, F., D. J. Armstrong, J. D. Cherayil, A. E. Hampel, and R. M. Bock. 1966. Cytokinin activity: localization in transfer RNA preparations. *Science* **154**: 1354–1356.

Skoog, F., H. Q. Hamzi, A. M. Szweykowska, N. J. Leonard, K. L. Carraway, T.

Fujii, J. P. Helgeson, and R. N. Loeppky. 1967. Cytokinins: structure/activity relationships. *Phytochemistry* **6**: 1169–1192.

Skoog, F. and C. O. Miller. 1957. Chemical regulation of growth and organ formation in plant tissues cultured *in vitro*. *Symp. Soc. Exp. Biol.* **11**: 118–131.

Skoog, F., R. Y. Schmitz, R. M. Bock, and S. M. Hecht. 1973. Cytokinin antagonists: synthesis and physiological effects of 7-substituted 3-methylpyrazolo-[4,3-*d*]pyrimidines. *Phytochemistry* **12**: 25–37.

Skoog, F., F. M. Strong, and C. O. Miller. 1965. Cytokinins. *Science* **148**: 532–533.

Sondheimer, E. and D. Tzou. 1971. The metabolism of hormones during seed germination and dormancy. II. The metabolism of 8-^{14}C-zeatin in bean axes. *Plant Physiol.* **47**: 516–520.

Sorokin, H. P. and K. V. Thimann. 1964. The histological basis for inhibition of axillary buds in *Pisum sativum* and the effects of auxins and kinetin on xylem development. *Protoplasma* **59**: 326–350.

Spiess, L. D. 1975. Comparative activity of isomers of zeatin and ribosyl-zeatin on *Funaria hygrometrica*. *Plant Physiol.* **55**: 583–585.

Steward, F. C. and A. D. Krikorian. 1971. *Plants, Chemicals and Growth*. Academic Press, New York.

Takegami, T. and K. Yoshida. 1975. Isolation and purification of cytokinin binding protein from tobacco leaves by affinity column chromatography. *Biochem. Biophys. Res. Commun.* **67**: 782–789.

Tavares, J. and H. Kende. 1970. The effect of 6-benzylaminopurine on protein metabolism in senescing corn leaves. *Phytochemistry* **9**: 1763–1770.

Thimann, K. V. 1977. *Hormone Action in the Whole Life of Plants*. University of Massachusetts Press, Amherst.

Thimann, K. V., T. Sachs, and K. N. Mathur. 1971. The mechanism of apical dominance in *Coleus*. *Physiol. Plantarum* **24**: 68–72.

Torrey, J. G. 1976. Root hormones and plant growth. *Ann. Rev. Plant Physiol.* **27**: 435–459.

Varner, J. E. and D. T. Ho. 1976. Hormones. In: Bonner, J. and J. E. Varner, eds. *Plant Biochemistry. 3rd ed.* Academic Press, New York. Pp. 713–770.

Vreman, H. J., F. Skoog, C. R. Frihart, and N. J. Leonard. 1972. Cytokinins in *Pisum* transfer ribonucleic acid. *Plant Physiol.* **49**: 848–851.

Walker, G. C., N. J. Leonard, D. J. Armstrong, N. Murai, and F. Skoog. 1974. The mode of incorporation of 6-benzylaminopurine into tobacco callus transfer ribonucleic acid. A double labeling determination. *Plant Physiol.* **54**: 737–743.

Wareing, P. F., R. Horgan, I. E. Henson, and W. Davis. 1977. Cytokinin relations in the whole plant. In: Pilet, P. E., ed. *Plant Growth Regulation*. Springer-Verlag, New York. Pp. 147–153.

Wareing, P. F. and I. D. J. Phillips. 1970. *The Control of Growth and Differentiation in Plants*. Pergamon Press, New York.

Wickson, M. and K. V. Thimann. 1958. The antagonism of auxin and kinetin in apical dominance. *Physiol. Plantarum* **11**: 62–74.

Zachau, H. G., D. Dütting, and H. Feldmann. 1966a. Nucleotide sequences of two serine-specific transfer ribonucleic acids. *Angew. Chem. Int. Edit.* **5**: 422.

Zachau, H. G., D. Dütting, and H. Feldmann. 1966b. The structures of two serine transfer ribonucleic acids. *Hoppe-Seylers Zeit. Physiol. Chem.* **347**: 212–235.

Chapter 5

Abscisic Acid and Related Compounds

Introduction

Regulation of the growth and development of plants is dependent upon the kinds and amounts of the various hormones and upon the changing sensitivity of the tissues to these substances throughout ontogeny. Yet the survival of plants as individual organisms and as populations and species over multiple generations depends also upon deceleration or cessation of growth and development of resistance to stress at critical stages in the life cycles. Coping with water stress and other environmentally imposed adverse conditions by relatively rapid physiological changes is a frequent and essential manifestation of the adaptation of terrestrial plants to their often highly changeable physical environments. Longer term and more gradually developing suspensions of growth, or deceleration to barely perceptible rates, called "dormancy," also are common features of the ontogeny of seed plants. Dormancy is a phenomenon of profound biological significance, because it provides a means by which plants are enabled to survive periods of environmental conditions which would be adverse or lethal to plants in an active state of growth. In the case of annual plants, dormancy of seeds is commonplace. In biennials and perennials, dormancy of the buds of established plants and of their seeds and storage organs (e.g., bulbs, corms, and tubers) are typical events in the life cycles.

Unfortunately, considerable confusion exists concerning the terminology relating to dormancy phenomena. But, by generally acceptable connotation, "dormancy" in the broad sense refers to any temporary suspension of active growth. Dormancy which is imposed by unfavorable physical environmental conditions such as low temperature or moisture stress, which often is immediately reversible upon the plant's again experiencing favorable conditions, is called "quiescence," "imposed dormancy," "temporary dormancy," or, sometimes, in the case of perennials, "summer dormancy." "True dormancy," "rest," or "innate dormancy" refer to a state of temporarily suspended growth which is due to internal conditions. In the case of true dormancy or rest, growth is suspended even under physical environmental conditions which apparently are favorable for growth. It is true dormancy or rest, hereafter called simply "dormancy," of buds and seeds, together with water stress, with which this chapter is concerned. Just as it is important to understand the environmental stimuli which evoke the onset and breaking of bud dormancy, and the stimuli involved in dormant seeds, it is of great interest to understand the changes in internal conditions which are causally related to the cycle of growth and dormancy. Especially prominent among these changes are alterations in the balance of hormones, particularly certain growth-stimulating hormones and one or more growth-inhibiting hormones.

Actually, a quite large number of chemically diverse growth-inhibiting substances have been identified in plants (Fig. 5-1), and some of these at least probably are very significant physiologically. However, presently abscisic acid (ABA) is the only inhibitor which is commonly regarded as a true hormone. Hence this chapter will be concerned chiefly with ABA and certain of its analogs. As F. T. Addicott and J. L. Lyon stated in a review in 1969, "Abscisic acid (ABA) is a plant hormone which now ranks in importance with the auxins, gibberellins, and cytokinins as a controlling factor in physiological processes."

History of Discovery

The discovery of abscisic acid (ABA) as a naturally occurring growth inhibitor occurred through independent investigations of different physiological phenomena in two different laboratories. F. T. Addicott et al. of the University of California, Davis since circa 1961 have been investigating natural substances which accelerate leaf abscission. Across the Atlantic, P. F. Wareing et al. of the University College of Wales at Aberystwyth have, for two decades or longer, been investigating natural inhibitors which appear to be causally related to bud dormancy in woody plants. By 1965 these two paths of research converged upon the discovery that a single hormone was involved in both. The history of how this came about is one of the most interesting stories in the physiology and biochemistry of plant hormones.

We can logically begin the story by noting the report by Liu and Carns

Figure 5-1. Representative inhibitors of the benzoic acid, cinnamic acid, lactone, and flavonoid types. (Redrawn from: *Plant Growth and Development*, 2nd ed., by A. C. Leopold and P. E. Kriedemann. Copyright © 1975 by McGraw-Hill Book Company. Used with permission of McGraw-Hill Book Company.)

in 1961 that they had isolated in crystalline form from dried, mature cotton burs a substance which they first called "abscisin" (later "abscisin I") which accelerates abscission of excised debladed cotton petioles. While Liu and Carns characterized some of the properties of "abscisin," they did not determine its structure, and it is assumed that the structure has not yet been determined. Anyway, interest in "abscisin I" waned until 1963, when Ohkuma et al. reported on the isolation in crystalline form of another abscission-accelerating substance which they termed "abscisin II" from young cotton fruit. They described some of the chemical and physiological properties of this new substance and proposed a tentative empirical formula of $C_{15}H_{20}O_4$. In the same year, 1963, C. F. Eagles and P. F. Wareing isolated chromatographically an inhibitor from leaves of *Betula pubescens* which could be applied to buds of growing seedlings and cause them to go dormant. They proposed the term "dormin" for this unknown substance and for "other substances which appear to function as endogenous dormancy inducers."

To expedite chemical identification of "dormin" Wareing sought assistance from chemists at Shell Research Ltd. at Milstead Laboratory in England. A short time later, in 1965, J. W. Cornforth, B. V. Milborrow, G. Rayback (all of Shell), and P. F. Wareing compared the properties of dormin (from *Acer pseudoplatanus* leaves) with those of abscisin II and concluded, from a comparison of molecular weights, infrared spectra, and melting points, that the two compounds are in fact identical. A molecular structure of abscisin II was proposed by Ohkuma et al. in 1965 (Fig. 5-2), and that this structure is valid for both "abscisin II" and "dormin" was confirmed by synthesis by Cornforth et al. later that same year (1965). By 1966 B. V. Milborrow and associates in England had perfected a sensitive

(R)–Abscisic acid

(S)–Abscisic acid

Figure 5-2. Structures of (R)- and (S)-abscisic acid. (S)-ABA is the naturally occurring enantiomorph.

spectropolarimetric method for identifying (S)-abscisic acid in plant extracts and for measurement of its concentration. In 1965 Ohkuma had already synthesized some analogs of abscisic acid and by 1966, (S)-abscisic acid had been identified from several species of seed plants.

It was only natural that some confusion began to develop over the terms "dormin" and "abscisin II;" hence, the investigators in Addicott's, Wareing's and the Shell groups got together at the Sixth International Conference on Plant Growth Substances, held in Ottawa in the summer of 1967, and suggested standardized terminology, as follows: "We now propose the term abscisic acid as a reasonable and useful compromise . . ." "We suggest that authors specify racemic abscisic acid . . . by calling it (RS)-abscisic acid, and that the naturally occurring enantiomorph be called (S)-abscisic acid when it is necessary to draw attention to the stereochemistry" (Fig. 5-2). "As an abbreviation of abscisic acid, we propose ABA." All these proposals were adopted promptly in the literature that followed. Except where noted otherwise, all further references to ABA in this chapter will be understood to be specifically to (S)-ABA.

Chemical Characterization

ABA is a sesquiterpenoid (15-C) compound which is related, by its biogenesis, to the monoterpenes, diterpenes (including GAs), carotenoids, and triterpenes. Available evidence indicates that the naturally occurring ABA is a single enantiomorph, specifically the dextrorotatory compound, (S)-ABA [or (+)-ABA]. (R)-ABA [or (−)-ABA], which accounts for about 50% of the racemic mixtures of ABA which are made synthetically, has biological activity equal to that of (S)-ABA in most cases, an exception being stomatal closure in which (R)-ABA reportedly is inactive. Several substances related to (S)-ABA have also been found to occur in plant tissues (Fig. 5-3). Likewise, a number of synthetic analogs have been synthesized (Fig. 5-4).

The structural features of ABA and related molecules which are essential for biological activity are that: (1) a free carboxyl group is essential, (2) the cyclohexane ring must contain a double bond in an α- or β-position, and (3) the configuration of the C-2 double bond must be *cis*. 2-*trans*-ABA occasionally shows activity, but this is due to its isomerization in light. Likewise, the activity of esters is due to the hydrolysis of the ester bond and formations of the free acid. The glucose ester of ABA (abscisyl-β-D-glucopyranoside) appears to be an effective storage form of ABA from which the free acid can readily be liberated by hydrolysis in vivo.

2-*trans*-Abscisic acid

(+)-Abscisyl-β-D-glucopyranoside

Phaseic acid

2-*trans*-Phaseic acid

4'-Dihydrophaseic acid

Theaspirone

Figure 5-3. Some naturally occurring substances related to ABA. (Reproduced, with permission, from the *Annual Review of Plant Physiology, Volume 20.* © 1969 by Annual Reviews Inc. From Addicott and Lyon, 1969.)

Biosynthesis and Other Features of Metabolism

There is very good evidence that ABA is synthesized from mevalonate via farnesyl pyrophosphate (Milborrow, 1974b; Robinson and Ryback, 1969), an intermediate in the isoprenoid biosynthetic pathway (Fig. 5-5). The derivation of GAs and ABA from farnesyl pyrophosphate is of particular interest from the standpoint of plant hormone physiology and biochemistry. In a later section we shall see that GAs and ABA generally have opposing physiological and biochemical effects. Thus it is of particular interest to elucidate the enzymic regulation that determines the relative rates of biosynthesis of GAs and ABA from the common intermediate, farnesyl pyrophosphate.

Structure	Physiological activity	Structure	Physiological activity
(R)–Abscisic acid	(−) Growth		(−) Growth
2, 4-*trans, cis*–Abscisic acid	(+) Abscission		(0) Abscission
	(+) Abscission		(0) Abscission
	(−) Growth		(+) Abscission (−) Growth

Figure 5-4. Structures and physiological activities of some synthetic substances related to (S)-ABA. (Reproduced, with permission, from the *Annual Review of Plant Physiology, Volume 20.* © 1969 by Annual Reviews Inc. From Addicott and Lyon, 1969.)

It appears that ABA, like at least some GA, biosynthesis occurs in chloroplasts. B. V. Milborrow (1974d) developed a cell-free enzyme system from lysed chloroplasts (and etioplasts) of avocado (*Persea gratissima*) fruit which biosynthesized abscisic acid from mevalonate. It cannot be said with certainty that all the ABA synthesis in avocado fruit occurs in plastids, but it is evident that at least the majority does. Milborrow obtained similar results with cell-free enzyme extracts of bean (*Phaseolus vulgaris*) and avocado leaves. Milborrow (1974a) did, however, present evidence that two pools of ABA can exist in ripening avocado fruit, raising questions about the compartmentation of this hormone.

A possible alternate pathway of biogenesis of ABA which has been described is from a carotenoid, specifically violaxanthin, via xanthoxin as an intermediate (Fig. 5-6). Xanthoxin has been demonstrated to be formed in vitro by high-intensity photoconversion of violaxanthin. However, this reaction has not been reported to occur in vivo. In other investigations,

Figure 5-5. Proposed pathway for the biosynthesis of (S)-abscisic acid from mevalonate.

Mevalonate

Carotenoids
(e.g., violaxanthin)

Photolysis or
lipoxygenase

Xanthoxin

ABA

Figure 5-6. Possible alternate pathway of biogenesis of (S)-ABA via a carotenoid and xanthoxin as intermediates. (Redrawn, with permission, from Varner, J. E. and D. T. Ho. 1976. Hormones. In: Bonner, J. and J. E. Varner, eds. *Plant Biochemistry. 3rd ed,* pp. 713–770. Copyright by Academic Press, Inc., New York.)

xanthoxin was observed to be formed by the lipoxygenase-catalyzed cleavage of violaxanthin. By whatever means it is formed, xanthoxin is of great interest. Xanthoxin does occur naturally in some plants, and growth inhibitory activity has been attributed to the compound. And in one very interesting case, there was indirect evidence that xanthoxin was rapidly converted to ABA in tomato shoots. The significance of this possible alternate pathway obviously deserves more research.

Three metabolic reactions have been demonstrated for ABA in plants. One is the easily reversible formation of a glucose ester (abscisyl-β-D-glucopyranoside) and the other two are the formations of the oxidation products phaseic acid and dihydrophaseic acid (Fig. 5-7), both of which are essentially inactive derivatives, via 6'-hydroxymethyl ABA.

One of the most dramatic features about ABA metabolism is a large and rapid increase in the endogenous levels when leaves of terrestrial higher plants are exposed to any of a variety of conditions which induce

Abscisyl–β–D–glucopyranoside

ABA

6'–Hydroxymethyl–ABA   Phaseic acid

4'–Dihydrophaseic acid

Figure 5-7. Metabolic reactions of (S)-ABA.

stress, including mineral starvation, flooding, injury, and drought (Fig. 5-8). ABA increases plant resistance to such stresses. The increase in ABA begins within minutes after the initiation of wilting, and may become 10-fold or greater than the concentration normally present in unwilted leaves. Furthermore, the high ABA levels decrease rapidly upon watering of water-stressed plants (Fig. 5-9). Phaseic acid and dihydrophaseic acid also increase in the leaves of water-stressed terrestrial plants, and in some cases—e.g., bean (*Phaseolus vulgaris*) leaves—the concentrations of these two oxidation products are higher than that of ABA, although this evidently is not true in all cases, e.g., castor bean (*Ricinus communis*). Results of investigations of beans and some other species indicate that the elevation of ABA levels in wilted leaves is due to de novo synthesis, rather than release from a conjugate such as ABA glucose ester. The key metabolic changes that occur in water-stressed bean leaves are: (1) water stress triggers an increased rate of ABA to phaseic and dihydrophaseic acids; (2) ABA concentration increases and becomes nearly constant as the increased rates of synthesis and metabolism become approximately equal; and (3) ABA concentration decreases as water potential increases and turgor is regained, due to decreased rates of ABA synthesis relative to metabolism (Harrison and Walton, 1975).

Milborrow (1974a) observed three patterns of regulation of ABA in different organs of avocado: (1) in the leaf there are relatively low normal levels of ABA (20 μg/kg) and on wilting there is an increase in biosynthesis until the overall concentration reaches approximately 2600 μg/kg; (2) in ripening fruit there are already high concentrations of ABA (4000–

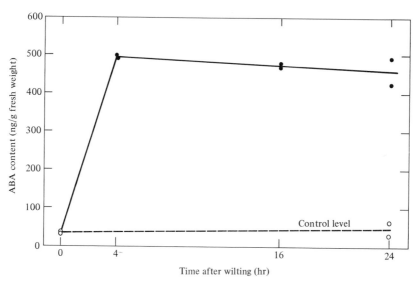

Figure 5-8. Increase in ABA content in wilting bean (*Phaseolus vulgaris*) leaves. Dashed line indicates level of ABA in control plants. (Redrawn, with permission, from Harrison and Walton, 1975.)

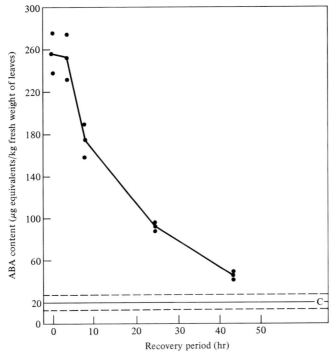

Figure 5-9. Decrease in ABA content of Brussels sprouts as they recover from wilt (44% leaf-water deficit). C = average ABA content of nonwilted plants. (Redrawn, with permission, from Wright and Hiron, 1972.)

6000 μg/kg), but in spite of this there is rapid biosynthesis, as determined by incorporation of mevalonate, and the ABA content and biosynthesis of fruit slices is unaffected by wilting; (3) in roots there are low ABA concentrations (27 μg/kg), and there is very little effect of wilting. Walton et al. (1976) obtained different results with roots of bean (*Phaseolus vulgaris*). These authors observed that ABA accumulates in water-deficient roots and decreases the resistance of roots to water uptake, thus further protecting the leaves against desiccation.

Zeevaart (1977) investigated the sites of (S)-ABA synthesis and metabolism in castor bean (*Ricinus communis*) by analyzing the levels of ABA, phaseic acid, and dihydrophaseic acid in the shoot tips, mature leaves, and phloem sap of water-stressed and nonstressed plants. He found that water stress caused increases in the concentrations of ABA, phaseic acid, and dihydrophaseic acid in all three materials analyzed. Dihydrophaseic acid, but not ABA, accumulated in shoot tips, indicating that ABA is metabolized rapidly, that the shoot tips are effective sinks for ABA translocated in the phloem sap, and that ABA is metabolized relatively rapidly in shoot tips. When young and mature leaves were excised and allowed to wilt, the ABA increased in both, demonstrating that leaves have the capacity to synthesize ABA at an early stage of development.

Figure 5-10. Structure of lunularic acid, an ABA-like substance found in some algae and liverworts.

Natural Occurrence of ABA

Originally ABA was found in the leaves of woody plants, *Acer pseudopla-tanus* and *Betula pubescens,* and in cotton fruits. Rapidly it was found to occur also in other seed plants. Thus far, in fact, the hormone has been found in all angiosperms and gymnosperms which have been investigated, as well as in at least one representative each from among ferns, horsetails and mosses, where concentrations occur in the range of 1 mμM to 1 μM. Some algae and liverworts reportedly contain a chemically related but different growth inhibitor called "lunularic acid" (Fig. 5-10), and so far it looks as if fungi and bacteria are devoid of ABA-like inhibitors.

Physiological Effects

Like all other plant hormones, ABA has multiple physiological effects, affecting growth and development literally throughout the ontogeny of seed plants. In this section, we shall be concerned with only four of the major physiological roles, namely in (1) stomatal regulation, (2) bud dormancy, (3) seed dormancy, and (4) abscission.

Stomatal Regulation

In recent years it has been discovered that ABA plays a major role in regulating closure of stomates, and thereby creating a protective mechanism toward water stress. The initial observation leading to this discovery was made as early as 1968 when it was noted that spray applications of exogenous ABA at low concentrations (e.g., 1 μM) could reduce transpiration, and by 1971 it was appreciated that this was due to closure of the stomates. Hence the possibility was raised that endogenous ABA might be involved in stomatal regulation and thus in the water relations of plants.

Onset of stomatal closure is quite rapid, occurring within approximately 3 to 9 minutes after the application of exogenous ABA to the cut bases of leaves of such plants as corn (*Zea Mays*), *Rumex obtusifolia,* and sugarbeet (*Beta vulgaris*). Moreover, the concentration of endogenous ABA increases sharply in water-stressed leaves during the lag period (10 to 15 minutes) prior to stomatal closure. ABA causes the guard cells to become "leaky" with respect to K^+, the guard cells lose K^+ and turgor, and the stomates close.

Actually the total process of stomatal regulation involves the concentrations of both ABA and CO_2 in the guard cells, and the response of the guard cells to either substance depends upon the concentration of the other. When, for example, water stress develops, ABA concentration in the guard cells increases, the guard cells lose K^+ and hence turgor, the stomates close—thus protecting the plant against desiccation—and CO_2 increases. Conversely, if the plant is watered and water stress subsides, ABA concentration in the guard cells decreases, K^+ and turgor are regained, CO_2 concentration decreases, and the stomates open—thus providing entry for CO_2 into the leaf and photosynthesis.

Bud Dormancy

The hypothesis that endogenous growth inhibitors may regulate bud dormancy was first put forth by T. Hemberg (1949a,b), who found that dormancy of buds of ash (*Fraxinus*) trees and potato tubers was correlated with a relatively high concentration of a growth inhibitor (or inhibitors) of unknown identity (Figs. 5-11 and 5-12). Kawase (1961) reported similar results for young birch (*Betula pubescens*) plants and confirmed further the effect of short days (10 hour) and long nights on inducing bud dormancy (Fig. 5-13).

Figure 5-11. Growth inhibitor content of bud scales of dormant (October) and nondormant (February) buds of ash (*Fraxinus*). (Redrawn from data of Hemberg, 1949a, in: *Plant Growth and Development,* 2nd ed., by A. C. Leopold and P. E. Kriedemann. Copyright © 1975 by McGraw-Hill Book Company. Used with permission of McGraw-Hill Book Company.)

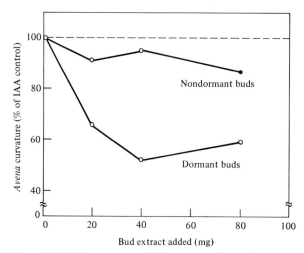

Figure 5-12. Growth inhibitor content of dormant and nondormant buds of potato tubers. Activity of extracts was measured as inhibition of auxin action in the *Avena* Coleoptile Curvature Test. (Redrawn from data of Hemberg, 1949b, in: *Plant Growth and Development* by A. C. Leopold. Copyright © 1964 by McGraw-Hill Book Company. Used with permission of McGraw-Hill Book Company.)

   Before discussing further the role of growth inhibitors in bud dormancy, it will be well to consider some generalizations about the phenomenon. Most, indeed probably all, perennial plants form resting buds at some stage in the annual cycle of growth. In temperate-zone woody plants dormant buds are formed typically in late summer or autumn, and the whole tree or shrub enters a dormant phase. Some temperate-zone woody species [e.g., birch (*Betula pubescens*) and larch (*Larix decidua*)] can be maintained in continuous growth for 18 months or longer under long-day conditions in a warm greenhouse but others cannot, and eventually all temperate-zone species set dormant buds even when kept under physical environmental conditions which evidently are optimum for growth.
   Resting buds also are formed by tropical species, but the dormancy behavior of these plants generally is very different from that of temperate-zone species. The latter are adapted to a physical environment in which critical environmental factors change predictably with the passing of the seasons. Both temperature and photoperiod change in repeating patterns, the latter with absolute precision at all points on Earth. In native tropical species, adapted to equatorial habitats where photoperiod changes very little (or not at all) and where adverse environmental conditions are more sporadic and typically the extremes are less severe, bud dormancy is under endogenous control (see Fig. 1-11). This behavior illustrates an important generalization about true dormancy, namely, that dormancy typically is not a response to adverse physical environmental conditions but rather an adaptive phenomenon which precedes and prepares plants

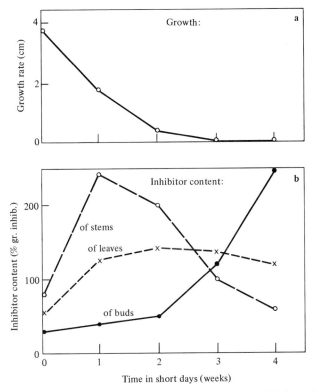

Figure 5-13. Induction of bud dormancy, **a**, by short days (10 hours) and long nights (14 hours) in birch (*Betula pubescens*), and correlated changes in growth inhibitor content, **b**, of leaves, stems, and buds. (Redrawn, with permission, from data of Kawase, 1961.)

for survival during exposure to adverse conditions. And, most interestingly, not all the shoots of a given tropical tree enter dormancy at the same time, so that both dormant and growing shoots occur on the tree at the same time. This behavior—endogenous regulation of dormancy—common in perennial species indigenous to tropical regions probably is the evolutionarily most primitive condition. The environmental conditioning and responses seen in species indigenous to higher and lower latitudes is believed to be an evolved, more advanced type of behavior which was essential to the evolution and northward and southward migrations of species from centers of origin of their progenitors in tropical areas.

One of the most important factors affecting and controlling the induction of bud dormancy in temperate-zone woody plants is day length. In the majority of woody plants studied so far (angiosperms and gymnosperms), long days promote vegetative growth and short days bring about the cessation of extension growth and the onset of bud dormancy. This photoperiodic induction of bud dormancy shows obvious parallels with photoperiodic phenomena in herbaceous plants. Phytochrome is the photoreceptor, and red/far-red reversibility in the "night break" effect has been demonstrated for seedlings of larch. The leaves commonly are the photoreceptors, but bud scales are also in some species (e.g., *Betula pubescens*).

There are, however, serious questions concerning the role of photoperiodism in the onset of dormancy of many woody plants, for a number of common cultivated fruit trees (e.g., *Pyrus, Malus, Prunus*) and certain other species (e.g., *Fraxinus excelsior* and other species in the *Oleaceae*) show very little response to day length. They may go dormant in summer when days still are relatively long. However, a word of caution is advisable about this. Through the efforts of horticulturists, many species of fruit trees are grown in temperate-zone areas which are substantially different from the areas to which the species are indigenous. It should not surprise us therefore to observe that species such as *Malus* (apple) and *Pyrus* (pear) species of Mediterranean origin do not behave like native species when grown in temperate regions. Also, even in species in which young trees exhibit conspicuous photoperiodic responses in the development of bud dormancy, older trees may not. Young trees of poplar and larch, for example, continue primary shoot growth into September or October, and go dormant in response to increasing night length. But older trees cease growth much earlier in the season, often in June, July, or August, when photoperiods are still long. What then controls the onset of bud dormancy in such plants? Adverse environmental conditions may play a role. That is, in many cases, the trees actually pass from a state of "predormancy" or "quiescence" into true dormancy or rest. Thus, although short-day conditions are not required for the initial formation of buds, they may play an important role in the ultimate development of deep dormancy.

Emergence from (breaking of) dormancy of perennial plants of temperate regions generally is conditioned by low temperature. Most woody plants need to be exposed to a period of winter chilling to overcome dormancy (e.g., in general, $1-10°$ C for 260 to > 1000 hours). Some species (e.g., *Acer pseudoplatanus* and *Betula pubescens*) can be induced to resume growth by transfer to long photoperiods, but this is not true for the majority of species of temperate-zone species. Although chilling is necessary to break dormancy of the buds, warm temperatures are necessary for the actual resumption of growth of the buds after the chilling period. Frequently the chilling requirements are met by January, but the buds fail to resume growth until later. They remain in a state of quiescence or postdormancy for awhile. Thus, the time of bud burst is normally determined by the return of warmer conditions.

Much of our present understanding of the hormonal regulation of bud dormancy stems from the extensive research by P. F. Wareing and his associates at the University College of Wales at Aberystwyth. They have used mainly *Betula pubescens* (birch) and *Acer pseudoplatanus* (sycamore maple). In both species dormancy can be induced by short photo-periods and long nights, and dormancy can be broken by long photo-periods and short nights (without cold treatment). They and others have shown that short-day conditions, which induce dormancy, result in in-creased levels of growth inhibitor in the leaves and buds of some species (Fig. 5-14). The inhibitory effect on the growth of buds exerted by leaves maintained under short-day conditions strongly suggests the translocation of inhibitor from the leaves to the buds. Furthermore, Wareing and his associates extracted inhibitor (putative ABA) from leaves of *Betula pube-scens* plants grown under short-day conditions, applied the extract to birch seedlings [also sycamore maple and black currant (*Ribes nigrum*)] maintained under long days, and the result was that the seedlings ceased growth and set resting buds complete with bud scales just as if they were caused to go dormant by photoperiodic induction. Moreover, they showed, as have numerous other investigators using other species, that the application of GA caused breaking of bud dormancy (whether dor-mancy had been naturally photoinduced or induced by the application of exogenous ABA).

The results obtained by Wareing and others with several species, chiefly *Betula pubescens* and *Acer pseudoplatanus,* indicate that ABA is causally involved in the regulation of the cycle of growth and dormancy. Thus P. F. Wareing put forth the concept that the annual cycle of bud growth and dormancy is regulated by a *balance* between endogenous growth inhibitors and GAs. It seems probable that induction of dormancy in at least some cases is brought about by high ABA and low GA levels,

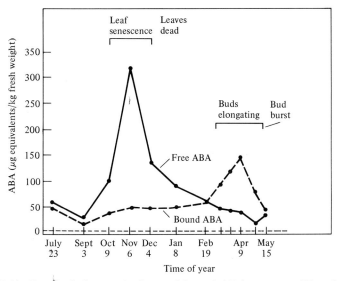

Figure 5-14. Seasonal changes in free and bound ABA content of Beech (*Fagus sylvatica*) buds in England. (Redrawn, with permission, from Wright, 1975.)

whereas the converse is true for the emergence from dormancy. Investigations based on analyses of xylem sap (e.g., peach and willow) and buds [e.g., apple, beech (Fig. 5-14) and birch] have revealed correlations between comparatively high ABA levels and the dormant period in several species of woody plants. In some cases the level of ABA in the xylem sap is 10 times higher during dormancy than during the growing period. Leaf buds do not open until well after ABA levels start to decrease. Moreover, it has been reported repeatedly that GA increases for many plant materials during chilling and long photoperiods. Thus it is very tempting to conclude that induction of bud dormancy by short days may be due mainly to high ABA levels, and that dormancy release may be determined primarily by increased GA levels.

But the conditions described above evidently do not apply in all cases; that is, even in all species of woody plants native to temperate regions. In some species—birch, sycamore maple, and red maple—it was reported that short-day treatment did not induce increases in endogenous ABA. In some other reported cases exogenous [^{14}C]ABA displayed little movement from the leaves to which it was applied, and direct applications of ABA to shoot tips stopped extension growth but did not cause normal set of terminal buds. In view of these results, some researchers have concluded that ABA may not actually play a causal role in bud dormancy. Obviously, although ABA and certainly GAs are strongly implicated, more research is needed to elucidate fully the hormonal regulation of bud dormancy in woody plants generally.

Seed Dormancy

Seeds of many species of angiosperms and gymnosperms will not germinate immediately after ripening even under conditions of moisture, temperature, and oxygen tension generally favorable for growth. Such dormancy, like dormancy of buds, is of obvious survival value to plants, since it tends to restrict germination to environmental circumstances suitable for seedling establishment and survival. Different seeds in one seed crop often vary in their degree of dormancy, causing germination of the crop to occur over an extended period of time and thus increasing the probability that at least part of them will germinate under conditions favorable for survival.

Seed dormancy results from particular conditions prevailing internally in the seed. Several causes of dormancy have been recognized, and seeds of numerous species exhibit two or more of them simultaneously. Among the many causes of seed dormancy which have been described are: (1) impermeability of seed coats to water and gases; (2) immaturity of the embryo; (3) need for "after-ripening" in dry storage; (4) mechanical resistance of seed coats; (5) presence of inhibitors found either in the seed coats, dry accessory structures, or, in the case of seeds contained in fleshy fruits, in the tissues surrounding the seeds; (6) special requirement for light or its absence; and (7) requirement for chilling in the hydrated condition.

The most common cause of seed dormancy in woody plants native to temperate regions is a requirement for chilling of the hydrated seeds. Such seeds commonly germinate promptly and uniformly only after they have become hydrated and exposed to low temperature (0 to 10°C) for a period of a few to several weeks. When these conditions are supplied artificially, the practice is referred to as stratification.

This dormancy mechanism, by far the most common seed dormancy mechanism in seeds of tree species, has obvious important survival value to many plants, in that it protects against winterkilling of seedlings. However, it is only relatively recently that we have begun to understand the effect of stratification and the dormant state at the biochemical level. Attention will focus mainly on two case histories of such studies. E. Sondheimer et al. (1968) investigated the ABA levels in relation to germination in seeds of two species of ash (*Fraxinus*). They used seeds of *Fraxinus americana,* a temperate-zone species, and *Fraxinus ornus,* a species of Mediterranean origin whose seeds do not display a stratification requirement. The results (Table 5-1) were extremely interesting. ABA was present in all tissues of both species, but the highest concentration was found in the seed and pericarp of dormant seeds of *Fraxinus americana.* During chilling (stratification) of seeds of *Fraxinus americana,* the ABA levels decreased 37% in the pericarp and 68% in the seed. And, of particular note, the ABA concentration in the seeds of *Fraxinus americana,* after chilling, was as low as the normally low level found in seeds of *Fraxinus ornus.* In the case of *Fraxinus americana* it seems clear that ABA in the seed has a regulatory role in germination, and that the

Table 5-1. (S)-Abscisic Acid Content of Seeds and Pericarps of *Fraxinus americana* and *Fraxinus ornus*[a]

Biological material	Germination as Σ 10[b]	Total (S)-ABA present in sample	
		μmols/kg	μg/1000 seeds or pericarps
F. americana			
Dormant seed	0–30	1.7	11
Chilled seed	700–950	0.6	3.3
Pericarp from dormant samara	—	2.8	24
Pericarp from chilled samara	—	1.8	15
F. ornus			
Seed	750	0.4	2.6
Pericarp	—	0.8	1.6

[a] Data from Sondheimer et al. (1968), with permission.
[b] Germination data are the sums of the daily percentages of germination for a 10-day period from the time of imbibition. Thus Σ 10 = 1000 would mean 100% germination during the first day.

most consequential result of stratification is to cause a decrease in endogenous ABA in the seed. Sondheimer et al., incidentally, concluded that it is unlikely that the ABA in the pericarp of *Fraxinus americana* seeds functions significantly in the regulation of dormancy.

Investigations by Sondheimer et al. showed further that other hormones besides ABA probably are involved in the regulation of the process of seed germination and seedling growth in *Fraxinus americana*. They noted, for example, that exogenous GA alone was effective in nullifying the effect of exogenous ABA on germination; however, leaf development and chlorophyll synthesis were less than normal. Exogenous kinetin, in combination with GA, promoted normal-appearing leaf development and chlorophyll synthesis. Kinetin by itself would not counteract the effect of ABA on germination. Exogenous ABA does, of course, prevent germination in nondormant seeds of both *Fraxinus americana* and *Fraxinus ornus*.

Another species with a stratification requirement which has been investigated extensively is *Corylus avellana* (known as hazelnut in Europe and the eastern U.S.A. and as filbert in the western U.S.A.). Exogenous GA is quite effective in causing nonstratified dormant seeds to germinate (Fig. 5-15), as is true generally for seeds with stratification requirements. In this case, chilling evoked an increase in endogenous GA, which occurred rapidly after stratified seeds were returned to a more normal ger-

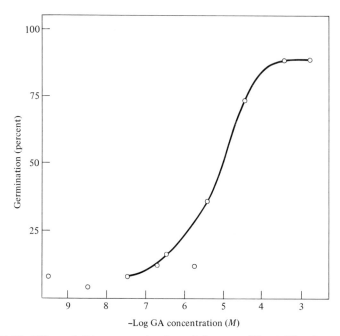

Figure 5-15. Effect of GA on germination of dormant filbert (*Corylus avellana*) seeds in the absence of stratification. (Redrawn, with permission, from Bradbeer and Pinfield, 1967.)

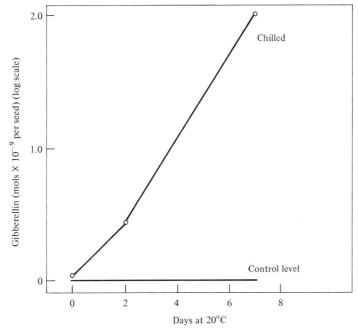

Figure 5-16. Increase in endogenous GA content of *Corylus avellana* seeds during the week following stratification for 28 days at 5°C. (Drawn from data of Ross and Bradbeer, 1968.)

mination temperature (Fig. 5-16). Moreover, Amo-1618 (an inhibitor of GA biosynthesis) counteracted the effect of chilling. An investigation of dormant *Corylus avellana* seeds by Bradbeer (1968) showed that while the dormant seeds contain inhibitor, ABA, if present at all, accounts for less than 1% of the total inhibitory activity. Hence, in the case of *Corylus avellana* it has been concluded that the breaking of dormancy by stratification may be attributed to the synthesis of GA.

Again, in the case of seeds having chilling requirements, as in the case of bud dormancy, the concept of balance between growth-promoting and growth-inhibiting hormones arises. In seeds evidently the balance can be altered in different ways during or immediately subsequent to stratification, either by a decrease of inhibitor (as in *Fraxinus americana*), increase in promoter (as in *Corylus avellana*), and perhaps by both such changes in other species.

Abscission

The role of ABA as a natural regulator of abscission of organs is not yet clear, although the hormone was discovered in part because of its capacity to cause abscission. L. A. Davis and F. T. Addicott measured ABA in cotton (*Gossypium hirsutum*) fruits by means of gas-liquid chromatography. They found that high levels of ABA occurred in direct correlation with abortion and abscission of young fruit, and with senescence and dehiscence of mature fruit. Young fruit that abscised late in the fruiting season contained twice as much ABA as young fruit that abscised early in the season. Hence, at least in the case of cotton fruits, ABA seems to be very much involved in the control of abscission. Its role in the natural control of leaf abscission is less clear, however. Commonly, relatively quite high levels of exogenous ABA are required to elicit leaf abscission. In the case of leaves abscission may be regulated mainly by other hormones.

State of Chromatin in Dormant Tissues and the Mechanism of Action of ABA

Dormant tissue is alive, it respires, and in some cases it grows very slowly. Yet clearly a dormant tissue or organ is idle; it does not grow normally, and obviously there is some mechanism which restricts growth.

Dorothy Tuan and James Bonner (1964) advanced the hypothesis ". . . that in the dormant cell the genetic material (the DNA) is completely, or nearly completely, repressed." According to this concept, of course, the processes of DNA transcription, the production of messenger RNA, and the production of enzymes and structural proteins essential for growth and metabolism, all are repressed. Tuan and Bonner conducted investigations to test the hypothesis. Specifically they compared: (1) the rates of RNA synthesis by dormant and nondormant buds of potato tubers; and (2) the capacities of the chromatin of such buds to support DNA-dependent RNA synthesis.

The results were highly revealing. Dormant buds possessed only a very limited capacity for both DNA and DNA-dependent RNA synthesis, as did the isolated chromatin from dormant buds, compared to buds in which dormancy was broken by treatment of the tubers with ethylene chlorohydrin. It was, therefore, concluded that the DNA of the buds of dormant potato tubers is largely in a repressed state, and that breaking of dormancy is accompanied by derepression of the genetic material.

Similar results were reported from investigations of dormant and nondormant seeds of *Corylus avellana* by Jarvis et al. (1968). They confirmed that dormant seeds require stratification and that exogenous GA can replace chilling. Using intact embryos and isolated chromatin, they found that GA first increased the amount of DNA available for transcription and secondarily caused an increase in RNA synthesis.

The molecular mechanism of action of ABA presently is unknown. However, there are two major biochemical effects of the hormone which are well documented. One is an alteration of plasma membranes, which is manifested in a change in bioelectric potential across the membranes and a leakiness and efflux of K⁺, which is involved in some actions of ABA, e.g., stomatal regulation which was discussed previously. The other major effect is inhibition of RNA and protein synthesis. ABA counteracts many of the effects of GA, and certain effects also of auxins and cytokinins. For example, ABA counteracts the effect of GA in the induction of hydrolases in barley aleurone layers (Fig. 3-31), and over a wide range of concentrations ABA apparently behaves as a competitive inhibitor of GA. The action of ABA on GA induction of hydrolases has largely been the basis for speculating that ABA may act specifically to inhibit DNA-dependent RNA synthesis.

References

Addicott, F. T. 1970. Plant hormones in the control of abscission. *Biol. Rev.* **45**: 485–524.

Addicott, F. T. and J. L. Lyon. 1969. Physiology of abscisic acid and related substances. *Ann. Rev. Plant Physiol.* **20**: 139–164.

Addicott, F. T., J. L. Lyon, K. Ohkuma, W. E. Thiessen, H. R. Carns, O. E. Smith, J. W. Cornforth, B. V. Milborrow, G. Ryback, and P. F. Wareing. 1968. Abscisic acid: a new name for abscisin II (dormin). *Science* **159**: 1493.

Alvim, R., E. W. Hewett, and P. F. Saunders. 1976. Seasonal variation in the hormone content of willow. I. Changes in abscisic acid content and cytokinin activity in the xylem sap. *Plant Physiol.* **57**: 474–476.

Alvim, R., S. Thomas, and P. F. Saunders. 1978. Seasonal variation in the hormone content of willow. II. Effect of photoperiod on growth and abscisic acid content of trees under field conditions. *Plant Physiol.* **62**: 779–780.

Audus, L. J. 1975. Geotropism in roots. In: Torrey, J. G. and D. T. Clarkson, eds. *The Development and Function of Roots.* Academic Press, New York. Pp. 327–363.

Bradbeer, J. W. 1968. Studies in seed dormancy. IV. The role of endogenous inhibitors and gibberellin in the dormancy and germination of *Corylus avellana* L. seeds. *Planta* **78**: 266–276.

Bradbeer, J. W. and N. J. Pinfield. 1967. Studies in seed dormancy. III. The effects of gibberellin on dormant seeds of *Corylus avellana* L. *New Phytol.* **66**: 515–523.

Chrispeels, M. J. and J. E. Varner. 1966. Inhibition of gibberellic acid induced formation of α-amylase by abscisin II. *Nature (London)* **212**: 1066–1067.

Chrispeels, M. J. and J. E. Varner. 1967. Hormonal control of enzyme synthesis: on the mode of action of gibberellic acid and abscisin in aleurone layers of barley. *Plant Physiol.* **42**: 1008–1016.

Colquhoun, A. J. and J. R. Hillman. 1975. Endogenous abscisic acid and the senescence of leaves of *Phaseolus vulgaris* L. *Z. Pflanzenphysiol.* **76**: 326–332.

Cornforth, J. W., B. V. Milborrow, and G. Ryback. 1966. Identification and estimation of (+)-abscisin II ('dormin') in plant extracts by spectropolarimetry. *Nature (London)* **210**: 627–628.

Cornforth, J. W., B. V. Milborrow, and G. Ryback. 1965. Synthesis of (±)-abscisin II. *Nature (London)* **206**: 715.

Cornforth, J. W., B. V. Milborrow, G. Ryback, and P. F. Wareing. 1965. Identity of sycamore 'dormin' with abscisin II. *Nature (London)* **205**: 1269–1270.

Davis, L. A. and F. T. Addicott. 1972. Abscisic acid: correlations with abscission and with development in the cotton fruit. *Plant Physiol.* **49**: 644–648.

Dennis, F. G., Jr. and L. J. Edgerton. 1961. The relationship between an inhibitor and rest in peach flower buds. *Proc. Am. Soc. Hort. Sci.* **77**: 107–116.

Downs, R. J. and H. A. Borthwick. 1956. Effects of photoperiod on growth of trees. *Bot. Gaz.* **117**: 310–326.

Eagles, C. F. and P. F. Wareing. 1963. Experimental induction of dormancy in *Betula pubescens. Nature (London)* **199**: 874–875.

Eagles, C. F. and P. F. Wareing. 1964. The role of growth substances in the regulation of bud dormancy. *Physiol. Plantarum* **17**: 697–709.

El-Antably, H. M. M. 1975. Redistribution of endogenous indoleacetic acid, abscisic acid and gibberellins in geotropically stimulated *Ribes nigrum* shoots. *Physiol. Plantarum* **34**: 167–170.

El-Antably, H. M. M., P. F. Wareing, and J. Hillman. 1967. Some physiological responses to d,l abscisin (dormin). *Planta* **73**: 74–90.

Evenari, M. 1949. Germination inhibitors. *Bot. Rev.* **15**: 153–194.

Evenari, M. 1957. The physiological action and biological importance of germination inhibitors. *Symp. Soc. Exp. Biol.* **11**: 21–43.

Firn, R. D., R. S. Burden, and H. F. Taylor. 1972. The detection and estimation of the growth inhibitor xanthoxin in plants. *Planta* **102**: 115–126.

Firn, R. D. and J. Friend. 1972. Enzymatic production of the plant growth inhibitor, xanthoxin. *Planta* **103**: 263–266.

Frankland, B. and P. F. Wareing. 1962. Changes in endogenous gibberellins in relation to chilling of dormant seeds. *Nature (London)* **194**: 313–314.

Greathouse, D. C., W. M. Laetsch, and B. O. Phinney. 1971. The shoot-growth rhythm of a tropical tree, *Theobroma cacao. Am. J. Bot.* **58**: 281–286.

Harrison, M. A. and P. F. Saunders. 1975. The abscisic acid content of dormant birch buds. *Planta* **123**: 291–298.

Harrison, M. A. and D. C. Walton. 1975. Abscisic acid metabolism in water-stressed bean leaves. *Plant Physiol.* **56**: 250–254.

Heide, O. M. 1974. Growth and dormancy in Norway spruce ecotypes (*Picea abies*). I. Interaction of photoperiod and temperature. *Physiol. Plantarum* **30**: 1–12.

Hemberg, T. 1949a. Growth inhibiting substances in buds of Fraxinus. *Physiol. Plantarum* **2**: 37–44.

Hemberg, T. 1949b. Significance of growth inhibiting substances and auxins for the rest-period of the potato tuber. *Physiol. Plantarum* **2**: 24–36.

Hemberg, T. 1958. The significance of the inhibitor β complex in the rest period of the potato tuber. *Physiol. Plantarum* **11**: 615–626.

Hemberg, T. 1961. Biogenous inhibitors. *Handbuch Pflanzenphysiol.* **14**: 1162–1183.

Hendershott, C. H. and L. F. Bailey. 1955. Growth inhibiting substances in extracts of dormant flower buds of peach. *Proc. Am. Soc. Hort. Sci.* **65**: 85–92.

Hendershott, C. H. and D. R. Walker. 1959. Identification of a growth inhibitor from extracts of dormant peach flower buds. *Science* **130**: 798–799.

Hocking, T. J. and J. R. Hillman. 1975. Studies on the role of abscisic acid in the initiation of bud dormancy in *Alnus glutinosa* and *Betula pubescens. Planta* **125**: 235–242.

Ingersoll, R. B. and O. E. Smith. 1970. Movement of (RS)-abscisic acid in the cotton explant. *Plant Physiol.* **45**: 576–578.

Jarvis, B. C., B. Frankland, and J. H. Cherry. 1968. Increased DNA template and RNA polymerase associated with the breaking of seed dormancy. *Plant Physiol.* **43**: 1734–1736.

Juniper, B. E. 1976. Geotropism. *Ann. Rev. Plant Physiol.* **27**: 385–406.

Kawase, M. 1961. Growth substances related to dormancy in *Betula*. *Proc. Am. Soc. Hort. Sci.* **78**: 532–544.

Ketring, D. L. 1973. Germination inhibitors. *Seed Sci. Tech.* **1**: 305–324.

Kahn, A. A. 1968. Inhibition of gibberellic acid-induced germination by abscisic acid and reversal by cytokinins. *Plant Physiol.* **43**: 1463–1465.

Kriedemann, P. E., B. R. Loveys, G. L. Fuller, and A. C. Leopold. 1972. Abscisic acid and stomatal regulation. *Plant Physiol.* **49**: 842–847.

Lenton, J. R., V. M. Perry, and P. F. Saunders. 1972. Endogenous abscisic acid in relation to photoperiodically induced bud dormancy. *Planta* **106**: 13–22.

Leopold, A. C. 1964. *Plant Growth and Development*. McGraw-Hill Book Company, New York.

Leopold, A. C. and P. E. Kriedemann. 1975. *Plant Growth and Development. 2nd ed.* McGraw-Hill Book Company, New York.

Lipe, W. N. and J. C. Crane. 1966. Dormancy regulation in peach seeds. *Science* **153**: 541–542.

Mayer, A. M. and Y. Shain. 1974. Control of seed germination. *Ann. Rev. Plant Physiol.* **25**: 167–193.

Milborrow, B. V. 1967. The identification of (+)-abscisin II [(+) dormin] in plants and measurement of its concentrations. *Planta* **76**: 93–113.

Milborrow, B. V. 1974a. Biosynthesis of abscisic acid and its regulation. In: *Plant Growth Substances 1973*. Hirokawa Publishing Company, Tokyo. Pp. 384–395.

Milborrow, B. V. 1974b. Chemistry and biochemistry of abscisic acid. In: Runeckles, V. C., E. Sondheimer, and D. C. Walton, eds. *The Chemistry and Biochemistry of Plant Hormones. Vol. 7. Recent Advances in Phytochemistry.* Academic Press, New York. Pp. 57–91.

Milborrow, B. V. 1974c. The chemistry and physiology of abscisic acid. *Ann. Rev. Plant Physiol.* **25**: 259–307.

Milborrow, B. V. 1974d. Biosynthesis of abscisic acid by a cell-free system. *Phytochemistry* **13**: 131–136.

Ohkuma, K. 1965. Synthesis of some analogs of abscisin II. *Agr. Biol. Chem.* **29**: 962–964.

Ohkuma, K., F. T. Addicott, O. E. Smith, and W. E. Thiessen. 1965. The structure of abscisin II. *Tetrahedron Lett.* **29**: 2529–2535.

Ohkuma, K., J. L. Lyon, F. T. Addicott, and O. E. Smith. 1963. Abscisin II, an abscission-accelerating substance from young cotton fruit. *Science* **142**: 1592–1593.

Pearson, J. A. and P. F. Wareing. 1969. Effect of abscisic acid on activity of chromatin. *Nature (London)* **221**: 672–673.

Perry, T. O. and H. Hellmers. 1973. Effects of abscisic acid on growth and dormancy of two races of red maple. *Bot. Gaz.* **134**: 283–289.

Phillips, I. D. J. and P. F. Wareing. 1958a. Effect of photoperiodic conditions on the level of growth inhibitors in *Acer pseudoplatanus*. *Naturwissenschaften* **45**: 317.

Phillips, I. D. J. and P. F. Wareing. 1958b. Studies in dormancy of sycamore. I. Seasonal changes in the growth-substance content of the shoot. *J. Exp. Bot.* **9**: 350–364.

Phillips, I. D. J. and P. F. Wareing. 1959. Studies in dormancy of sycamore. II. The effect of daylength on the natural growth-inhibitor content of the shoot. *J. Exp. Bot.* **10**: 504–514.

Pilet, P. E. 1977. Growth inhibitors in growing and geostimulated maize roots. In: Pilet, P. E., ed. *Plant Growth Regulation*. Springer-Verlag, New York. Pp. 115–128.

Pryce, R. J. 1972. The occurrence of lunularic and abscisic acids in plants. *Phytochemistry* **11**: 1759–1761.

Robinson, D. R. and G. Ryback. 1969. Incorporation of tritium from [(4R)-4-³H]mevalonate into abscisic acid. *Biochem. J.* **113**: 895–897.

Robinson, P. M., P. F. Wareing, and T. H. Thomas. 1963. Isolation of the inhibitor varying with photoperiod in *Acer pseudoplatanus*. *Nature (London)* **199**: 875–876.

Robinson, P. M. and P. F. Wareing. 1964. Chemical nature and biological properties of the inhibitor varying with photoperiod in sycamore (*Acer pseudoplatanus*). *Physiol. Plantarum* **17**: 314–323.

Ross, J. D. and J. W. Bradbeer. 1968. Concentrations of gibberellin in chilled hazel seeds. *Nature (London)* **220**: 85–86.

Samish, R. M. 1954. Dormancy in woody plants. *Ann. Rev. Plant Physiol.* **5**: 183–204.

Schrempf, M., R. L. Satter, and A. W. Galston. 1976. Potassium-linked chloride fluxes during rhythmic leaf movement of *Albizzia julibrissin*. *Plant Physiol.* **58**: 190–192.

Sibaoka, T. 1969. Physiology of rapid movements in higher plants. *Ann. Rev. Plant Physiol.* **20**: 165–184.

Sondheimer, E. and E. C. Galson. 1966. Effects of abscisin II and other plant growth substances on germination of seeds with stratification requirements. *Plant Physiol.* **41**: 1397–1398.

Sondheimer, E., D. S. Tzou, and E. C. Galson. 1968. Abscisic acid levels and seed dormancy. *Plant Physiol.* **43**: 1443–1447.

Sondheimer, E. and D. C. Walton. 1970. Structure-activity correlations with compounds related to abscisic acid. *Plant Physiol.* **45**: 244–248.

Steadman, J. R. and L. Sequeira. 1970. Abscisic acid in tobacco plants. Tentative identification and its relation to stunting induced by *Pseudomonas solanacearum*. *Plant Physiol.* **45**: 691–697.

Taylor, H. F. and T. A. Smith. 1967. Production of plant growth inhibitors from xanthophylls: a possible source of dormin. *Nature (London)* **215**: 1513–1514.

Thomas, T. H., P. F. Wareing, and P. M. Robinson. 1965. Action of sycamore 'dormin' as a gibberellin antagonist. *Nature (London)* **205**: 1270–1272.

Tuan, D. Y. H. and J. Bonner. 1964. Dormancy associated with repression of genetic activity. *Plant Physiol.* **39**: 768–772.

Varner, J. E. and D. T. Ho. 1976. Hormones. In: Bonner, J. and J. E. Varner, eds. *Plant Biochemistry*. 3rd ed. Academic Press, New York. Pp. 713–770.

Vegis, A. 1964. Dormancy in higher plants. *Ann. Rev. Plant Physiol.* **15**: 185–224.

Walton, D. C., M. A. Harrison, and P. Coté. 1976. The effects of water stress on abscisic-acid levels and metabolism in roots of *Phaseolus vulgaris* L. and other plants. *Planta* **131**: 141–144.

Walton, D. C. and E. Sondheimer. 1972. Metabolism of 2-^{14}C($\pm$)-abscisic acid in excised bean axes. *Plant Physiol.* **49**: 285–289.

Walton, D. C., G. S. Soofi, and E. Sondheimer. 1970. The effects of abscisic acid on growth and nucleic acid synthesis in excised embryonic bean axes. *Plant Physiol.* **45**: 37–40.

Wareing, P. F. 1954. Growth studies in woody species. VI: The locus of photoperiodic perception in relation to dormancy. *Physiol. Plantarum* **7**: 261–277.

Wareing, P. F. 1956. Photoperiodism in woody plants. *Ann. Rev. Plant Physiol.* **7**: 191–214.

Wareing, P. F. 1969. The control of bud dormancy in seed plants. *Symp. Soc. Exp. Biol.* **23**: 241–262.

Wareing, P. F. and D. L. Roberts. 1956. Photoperiodic control of cambial activity in *Robinia pseudacacia* L. *New Phytol.* **55**: 356–366.

Wareing, P. F. and P. F. Saunders. 1971. Hormones and dormancy. *Ann. Rev. Plant Physiol.* **22**: 261–288.

Wilkins, H. and R. L. Wain. 1975. Abscisic acid and the response of the roots of *Zea mays* L. seedlings to gravity. *Planta* **126**: 19–23.

Wright, S. T. C. 1975. Seasonal changes in the levels of free and bound abscisic acid in blackcurrent (*Ribes nigrum*) buds and beech (*Fagus sylvatica*) buds. *J. Exp. Bot.* **26**: 161–174.

Wright, S. T. C. and R. W. P. Hiron. 1972. The accumulation of abscisic acid in plants during wilting and under other stress conditions. In: Carr, D. J., ed. *Plant Growth Substances 1970.* Springer-Verlag, New York. Pp. 291–298.

Zeevaart, J. A. D. 1977. Sites of abscisic acid synthesis and metabolism in *Ricinus communis* L. *Plant Physiol.* **59**: 788–791.

Chapter 6

Ethylene

Historical Background

General recognition of ethylene, a simple hydrocarbon gas (C_2H_4), as a plant hormone has come about only relatively recently, although it has been known for more than three-fourths of a century that the gas has numerous interesting effects on growth and development. D. N. Neljubow, a Russian physiologist, evidently was the first scientist to write about ethylene action on plants. In 1901 he recorded that ethylene, which he identified as a component of illuminating gas, causes the classical "triple response" of etiolated pea seedlings. This response includes inhibition of stem elongation, increase in radial expansion (swelling) of stems, and a horizontal orientation of stems to gravity (see Fig. 6-12). Some years later (1910) H. H. Cousins advised the Jamaican government that oranges and bananas should not be stored together on ships because some unidentified volatile agent would cause the bananas to ripen prematurely.

Several workers reported in the years between about 1917 and 1937 that ethylene stimulates fruits to ripen. During the years from about 1933 to 1937, several investigators showed further that ethylene is produced by many plant materials, especially fleshy fruits. Indeed, the observations of several responses of plants to exogenous ethylene and the observation that a variety of vegetative tissues as well as fleshy fruits produce ethylene led W. Crocker, A. E. Hitchcock, and P. W. Zimmerman of the Boyce Thompson Institute for Plant Research in New York to suggest in 1935 that ethylene might be an endogenous growth regulator, and that it be regarded as a ripening hormone. Unfortunately, however, interest in ethylene among plant physiologists did not become intense or even widespread until nearly 24 years later. In good part the reason for comparatively little attention to ethylene was the lack of availability of techniques to analyze the relatively small amounts of the gas in plant tissues generally. About 1959 interest in the biochemistry and physiology intensified when Stanley P. Burg, Harlan K. Pratt, and others began applying the newly developed technique of flame ionization gas chromatography to ethylene analyses. This advanced the limit of analysis nearly a million-fold over the older manometric techniques which had been used previously. By flame ionization gas chromatography, it is possible to measure a fraction of a part per billion, or as little as 10^{-3} pmols, in a small gas sample.

Still the general acceptance of ethylene as a hormone was rather slow to develop. For, as Burg (1962) observed, ". . . it (ethylene) would be unique in that it is a vapor at physiological temperatures so that its production, accumulation, transport, and function would involve special problems not encountered in other areas of hormone physiology." Unlike some other hormones, ethylene apparently does not undergo directed transport. Hence it somehow accomplishes a hormonal, integrative function by diffusing rapidly through tissues. But by 1969, Pratt and Goeschl could say that, "With very sensitive instruments and very careful technique, it has become possible to show that ethylene is an endogenous growth regulator in plants. . . ."

Resurgence and great expansion of interest in ethylene biochemistry and physiology occurred in the late 1950s, and this has been a very active and productive field of research ever since. Ethylene now is understood to be a natural product of plant metabolism, which is produced by healthy, as well as senescent and diseased, tissues and which exerts regulatory control or influence over plant growth and development generally throughout ontogeny. Its interactions with other kinds of hormones are numerous and complex. Among the most diverse physiological effects which have been reported are:

1. Stimulation of ripening of fleshy fruits.
2. Stimulation of leaf abscission.
3. "Triple response" of etiolated legume seedlings—reduced stem elongation, radial swelling of stems, and ageotropism or diageotropism of stems.
4. Inhibition of leaf and terminal bud expansion in etiolated seedlings.
5. Tightening of the epicotyl or hypocotyl hook of etiolated dicot seedlings.
6. Inhibition of root growth.
7. Increase in membrane permeability.
8. Stimulation of adventitious root formation.
9. Stimulation of flowering in pineapple.
10. Inhibition of lateral bud development.
11. Causes various types of flowers to fade.

12. Interference with polar auxin transport.
13. Causes epinasty of leaves.
14. Participation in normal root geotropism.

Ethylene and Fruit Ripening

For many years ethylene physiology remained primarily the province of those interested in fruit physiology. Several practical applications developed from this research, among them being the process of "gas storage" or "controlled atmosphere" (CA) storage for the preservation of fruits, which was developed by Kidd and West in 1933. It had been determined that oxygen is necessary for ethylene action and that carbon dioxide inhibits action of the gas. Basically the process of "gas storage" consists of storing fruits in an atmosphere rich in carbon dioxide (5 to 10%), low in oxygen (1 to 3%), and with as little ethylene as possible. Gas storage is achieved commerically by storing fruits in an airtight room at low temperature. The oxygen is naturally depleted and the carbon dioxide increased by respiration. Ethylene produced by the fruits is absorbed on brominated charcoal filters. The effects of the controlled atmosphere are explainable on the grounds that oxygen is essential for ethylene action (Fig. 6-1) and carbon dioxide is a competitive inhibitor of ethylene (Fig. 6-2).

Numerous changes are associated with the ripening of fleshy fruits, including: (1) softening, (2) hydrolysis of storage materials, (3) changes in

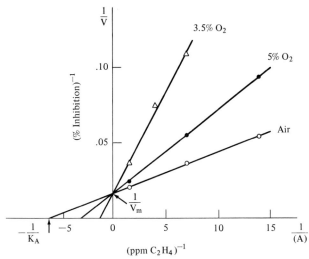

Figure 6-1. Lineweaver–Burk double reciprocal plot (percentage inhibition^{-1} versus concentration of ethylene^{-1}) at various levels of O_2, illustrating positive effect of O_2 on ethylene action. Data are based on inhibition of growth of pea stem sections by ethylene. (Redrawn, with permission, from Burg and Burg, 1967a.)

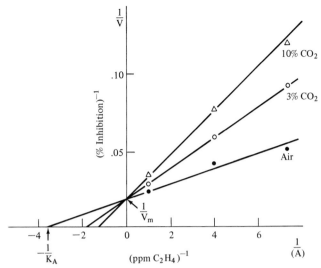

Figure 6-2. Lineweaver–Burk double reciprocal plot (percentage inhibition^{-1} versus concentration of ethylene^{-1}) at various CO_2 concentrations, showing competitive inhibition of ethylene action by CO_2. Data are based on inhibition of growth of pea root sections by ethylene. (Redrawn, with permission, from Chadwick and Burg, 1967.)

pigmentation and flavor, and (4) changes in respiration rate. The fleshy fruit is more complex in its senescence than other plant organs, and all of the changes which occur do not represent deterioration, but in a real sense fruit ripening may be regarded as a special case of organ senescence. In many kinds of fleshy fruits, there are marked changes in respiration rate after the fruits mature (reach final size). In some species there is a decrease in respiration rate in the mature fruit, followed by a large increase during the time of ripening (the respiratory climacteric), and a final decrease in respiration rate as the fruit enters a senescent decline (Fig. 6-3). Examples of fruits which show climacteric are avocado, banana, pear, apple, mango. Not all fleshy fruits exhibit a respiratory climacteric, however. There are in fact two types of nonclimacteric fruits: (1) those that show a steady respiration rate during ripening (e.g., orange, lemon, fig); and (2) those that show a decline in respiration rate during the ripening period (e.g., pepper).

The induction of ripening and senescent changes induced in many kinds of fruits by ethylene has been repeatedly demonstrated over the past 50 years. However, the role of ethylene in the metabolism of fruits, and particularly in the respiratory climacteric, has been vigorously debated. As recently as 1960, some authors have argued that ethylene is a by-product rather than the cause of ripening. Their argument was based on their findings, largely by relatively insensitive manometric techniques, that ethylene is not always present in large enough quantity prior to the climacteric to stimulate ripening. On the other hand, Stanley P. Burg of

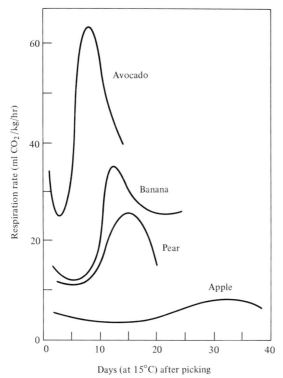

Figure 6-3. The course of the climacteric rise in fruit respiration during the postharvest period. (Reproduced, with permission, from the *Annual Review of Plant Physiology, Volume 1.*© 1950 by Annual Reviews Inc. From Biale, 1950.)

the University of Miami, Miami, Florida, has, in the years since circa 1959, presented what many regard as convincing evidence that ethylene is, in fact, the hormone which initiates fruit ripening. Burg has utilized consistently the technique of flame ionization gas chromatography in his ethylene analyses. He has shown that even fruits which do not normally exhibit a respiratory climacteric can be induced to do so by applying ethylene (e.g., lemon and orange). Moreover, Burg and others have shown that for the climacteric fruits studied, there is a rise in the intercellular content of ethylene to a level which stimulates these fruits to ripen— well in advance of the respiratory climacteric (Fig. 6-4). But actually there is considerable variation among species, and H. K. Pratt and J. D. Goeschl (1969) stated that fleshy fruits fall into several classes as regards their ripening response to ethylene. In most fleshy fruits, the onset of ripening is preceded by an increase in ethylene production. In a few species (e.g., avocado and mango), significant concentrations of ethylene may be present some time before ripening, but the response to ethylene is inhibited until after the fruit is harvested. In apple, it appears that a step in ethylene synthesis may be inhibited by a product from the parent tree. When detached from the tree the inhibitor presumably disappears and the rate of ethylene production rises. In another group, including certain kinds of banana, a presumably effective ethylene concentration may be

Figure 6-4. Relationship between endogenous ethylene content and respiration rate in banana. (Redrawn, with permission, from Burg, S. P. and E. A. Burg. 1965b. Ethylene action and the ripening of fruits. *Science* **148**: 1190–1196, 28 May 1965. Copyright 1965 by the American Association for the Advancement of Science.)

present in the unripe fruit, but the fruit is insensitive to that concentration at that stage of development. As the fruit matures, it becomes more sensitive, and ripening ensues when sensitivity matches the endogenous ethylene content. In several annual crops (e.g., tomato and melons) ethylene production, ripening, and senescence occur at about the same age from anthesis, regardless of age at harvest. It appears that, in these fruits, the basic ethylene-forming mechanism exists and operates at a low level, but is not fully activated until the fruit reaches a critical physiological age.

Evidently ethylene causes two major types of biochemical effects in the process of fruit ripening. The oldest evidence is for changes in membrane permeability, leading to altered compartmentation and release of enzymes associated with such ripening processes as respiration and breakdown of acids and wall constituents. Permeability changes reportedly do occur in cellular membranes preceding or during ripening, as indicated by a leakage of solutes from the cells and also an increase in free space and filling of intercellular spaces with juice. Yet whether, in fact, ethylene causes permeability changes remains an unresolved question. Some investigators have argued that if there is a direct effect of ethylene on one or more membranes, it is more likely to be a specific effect involving only certain solutes rather than a general increase in permeability. Meanwhile, as regards fruit ripening, even though increased membrane permeability may not be the direct cause, such permeability changes may make possible the metabolic processes which collectively comprise ripening. And, in any event, of equal or perhaps greater importance are the biochemical changes correlated with the observed increases in protein (enzyme) synthesis that occur during the climacteric.

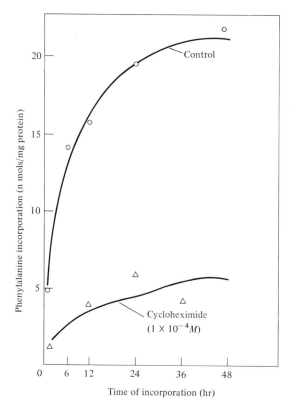

Figure 6-5. Effect of cyclo-heximide on incorporation of [^{14}C]phenylalanine into protein in midclimacteric pears. (Redrawn with modifications, by permission, from Frenkel et al., 1968.)

Clearly there is an increase in protein content during the climacteric (Fig. 6-5). It is suggested by available evidence that ripening processes are correlated with the formation of enzymes which catalyze the biochemical changes occurring during ripening. If mature fruit tissues are treated with inhibitors of protein synthesis, ripening is prevented parallel with the reduction in protein synthesis (Fig. 6-6). Ethylene synthesis itself apparently is dependent upon protein synthesis at the early climacteric stage, but less so once the climacteric has progressed.

There is no doubt that fruit tissues respond (directly or indirectly) to ethylene by exhibiting increases in the activities of enzymes that catalyze ripening reactions—so do other target tissues respond likewise to ethylene—and in many cases the increases in enzyme activity probably are the result of de novo enzyme synthesis, rather than activation of preexisting enzyme. But a yet unanswered question is whether ethylene acts directly to evoke new enzyme production. Interpretation of results with inhibitors of RNA and protein synthesis leaves some uncertainty, because it could be merely that RNA and protein synthesis are essential to maintain the cells in a state competent to respond to ethylene. Moreover, there are some responses to ethylene, besides fruit ripening, which occur apparently in too brief a time to allow for the direct involvement of RNA and

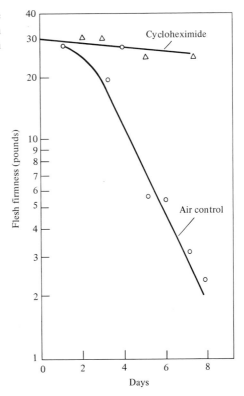

Figure 6-6. Effect of cycloheximide on ripening in pears. (Redrawn with modifications, by permission, from Frenkel et al., 1968.)

protein synthesis. Obviously, additional research is required to elucidate the primary action of ethylene in responsive tissues.

Interaction between Auxin and Ethylene

Zimmerman and Wilcoxon first reported in 1935 that exogenous auxin stimulates ethylene production in several plants. Since about 1964, several workers have confirmed and extended that observation. Physiological concentrations of IAA and other auxins stimulate ethylene formation in the roots, stems, flowers, fruits, and leaves of all plants which have been examined. By circa 1966 an hypothesis had developed which rapidly gained wide support, which attributes many of the effects of auxin to its influence on ethylene production.

The response of plant tissues to auxin varies with the concentration of auxin applied. In most cases, an optimum auxin concentration is found, with lower concentrations being promotive and higher concentrations inhibitory, or at least supraoptimal (Fig. 6-7). The reason for the growth inhibition (or submaximal response) was for a long time unknown. But since

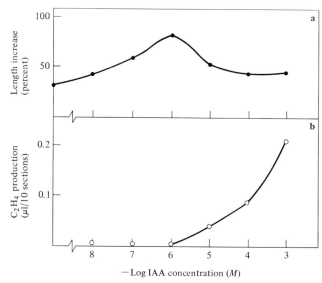

Figure 6-7. Growth response, **a**, and ethylene production, **b**, in etiolated pea stem sections treated with different concentrations of auxin during 18 hours in sealed flasks. (Redrawn with modifications, by permission, from Burg and Burg, 1966b.)

circa 1966 Stanley P. Burg and others have shown that in some systems auxin itself only promotes growth, and that auxin per se is never inhibitory. What happens is that at certain critical concentrations of auxin, which are different for each tissue, the production of ethylene is induced, and it is the ethylene that directly inhibits growth (Fig. 6-7). Thus, in some cases the inhibition of growth, e.g., elongation of etiolated pea stem sections (Fig. 6-7), etiolated sunflower stem sections, and roots (Fig. 6-8) by high concentrations of IAA evidently can be completely accounted for by auxin-stimulated ethylene production. However, the situation with some materials, e.g., sections of etiolated corn and oat coleoptiles, is different. Auxin stimulates these tissues to evolve ethylene, but the production does not continually increase as the IAA concentration is raised. Instead, ethylene formation and growth both reach their highest rates at about 10^{-5} M IAA, and they both decline as the auxin concentration is increased further. Hence it seems that auxin-induced growth inhibitions may occur for basically different reasons in coleoptiles and seedlings such as sunflower and pea. Sections of light-grown pea stems also behave quite differently than etiolated stem sections. Growth rate does not saturate at any concentration of IAA; that is, there is no sharp optimum (Fig. 6-9). Green stem sections are less responsive to ethylene than etiolated seedlings and produce ethylene at a slower rate in the presence of 10^{-5} or 10^{-4} M IAA. Consequently, even though green tissue swells when it responds to its own ethylene, the inhibition of elongation due to auxin-induced ethylene formation is never sufficient to offset the auxin-induced increase in length.

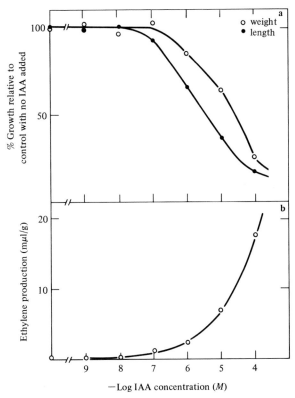

Figure 6-8. Growth inhibition, **a**, and ethylene production, **b**, in apical segments of pea roots treated with different concentrations of auxin. (Redrawn, with permission, from Chadwick and Burg, 1967.

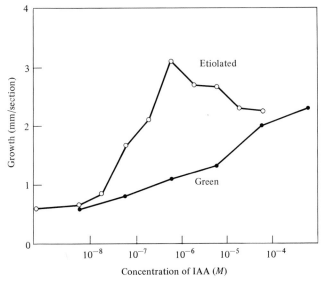

Figure 6-9. Growth responses of etiolated and light-grown pea stem sections to different concentrations of IAA. (Redrawn from data of Galston and Kaur, 1961.)

Inhibition of Root Growth and Role
in Root Geotropism

Previously it was pointed out that roots are extremely sensitive to auxin, inhibition of root sections occurring at concentrations of auxin which promote growth of stem sections, and that auxin inhibition of root growth was directly correlated with auxin-induced production of ethylene. Evidence reported by Chadwick and Burg (1970) strongly indicates that, indeed, auxin-induced inhibition of root growth is mediated entirely by auxin-stimulated formation of ethylene in vivo, as well as in root sections incubated with auxin in vitro. Moreover, ethylene—rather than auxin—may be the agent which directly mediates geotropism of roots—but not of stems.

According to the Cholodny–Went theory of geotropism, geotropism results from asymmetric distribution of endogenous auxin under the stimulus of gravity, such that the ratio of concentrations of endogenous auxin in the geostimulated organs becomes approximately 30:70 in the upper and lower halves, respectively. Roots being more sensitive to auxin than stems, the additional increment of auxin on the lower side of the root is sufficient to cause inhibition of growth on the lower side relative to the upper. Thus geostimulated roots curve downward when a growth inhibitory concentration of auxin accumulates in their lower side. Conversely, in geostimulated stems only differential growth promotion results from the asymmetric distribution of endogenous auxin.

The involvement of ethylene in root geotropism helps to overcome several objections which have been raised against application of the Cholodny–Went theory to roots. More than two decades ago Audus and Brown noted that during geotropism of roots growth of the upper side is, in fact, first promoted then inhibited. They proposed that auxin per se is not the active inhibitor, but rather an inhibitor which moves from the lower side to the upper side of a geostimulated root. Ethylene fits this concept well, since the ethylene produced in the lower side, as a result of unequal distribution of auxin, would be expected eventually to diffuse to the upper side to some extent, and thus account for the slight inhibition of growth which subsequently occurs there. Moreover, results of investigations of ethylene transport by Burg and associates help to eliminate a second objection to the Cholodny–Went theory, namely, that the observed 30:70 distribution of auxin in horizontally oriented roots is insufficient to account for the amount of growth inhibition on the lower side of the root, since there are indications that low concentrations of applied IAA cause very extensive initial inhibition, followed by some recovery, and that the distribution between the upper and lower sides can approach 10:90 rather than the 30:70 previously reported for geostimulated stems.

But it must not be too hastily concluded that auxin-induced ethylene is

the only growth-inhibiting substance involved in root geotropism, for the
Cholodny–Went theory has been severely challenged on two major
grounds: (1) several investigators have not found consistent differences in
the IAA concentrations in the upper and lower halves of geostimulated
roots; and (2) there is mounting evidence for the involvement of one or
more growth inhibitors, one of which is ABA. ABA is found in the root
cap, it can be shown to be redistributed in the root cap, it moves basipetally in the root extension zone, and it does in fact inhibit growth in the
extension zone. Available evidence shows that the inhibitor is produced
only in the root cap and is laterally transported only in the root cap and
not in the elongating zone. In view of these collective findings, the role of
IAA, hence also IAA-induced ethylene, presently is uncertain.

Role in Emergence of Dicot Seedlings

Protection of the tender shoot tip from injury during seed germination and
emergence is an important feature of all seed plants. In species that exhibit hypogean germination (e.g., *Pisum sativum,* garden pea) (Fig. 6-10),
the cotyledons remain below ground (hypocotyl does not elongate) and
the emergent plumule is arched or recurved near the apex, so that the
shoot tip is pushed to the soil surface in a protected position. The common garden bean (*Phaseolus vulgaris*), on the other hand, displays epi-
gean germination, according to which the hypocotyl elongates, forming a
hook or arch, and the shoot tip is protected as it is pulled upward by the
elongating hypocotyl. In both cases, establishment of symmetrical growth
in the elongating shoot causes the epicotyl (or plumular) or hypocotyl
hook to be straightened out as the seedlings emerge above ground.

It turns out that this is yet another of the many aspects of growth and
development in which ethylene has an important hormonal role. In the
case of the etiolated pea seedling, for example, there is a localized production of ethylene in the plumule and plumular hook portion of the epicotyl at a rate of approximately 6 μl kg^{-1} hour^{-1}. This rate is sufficient to
yield physiologically active concentrations of ethylene within the tissue.
The localized production of ethylene, in fact, is responsible for formation
and maintenance of the hook in the etiolated seedling. As the etiolated
seedling emerges naturally from soil, or is experimentally irradiated with
white light, there is a transient decrease in ethylene production in the
hook region and establishment of symmetrical growth about the shoot
(Fig. 6-11). Moreover, as the etiolated tissue greens it becomes less sensitive to the ethylene that is produced. In determining the action spectrum
for the plumular expansion, red light (in the region of 660 nm) was most
effective. Moreover, far-red irradiation (approximately 730 nm) immedi-

Figure 6-10. Light-grown, **a**, and etiolated, **b**, seedlings of pea (*Pisum sativum*). (Adapted from *The Living Plant* by Peter Martin Ray. Copyright 1963 by Holt, Rinehart & Winston, Inc. Redrawn by permission of Holt, Rinehart & Winston.)

ately following the red decreased the red effect to the level achieved by the far-red alone. These results suggested that the ethylene production mechansim is mediated by phytochrome (Chapter 7) and that ethylene acts as a hormone in the phytochrome control of plumular expansion. This explanation applies equally well to seedlings that develop hypocotyl hooks, such as bean.

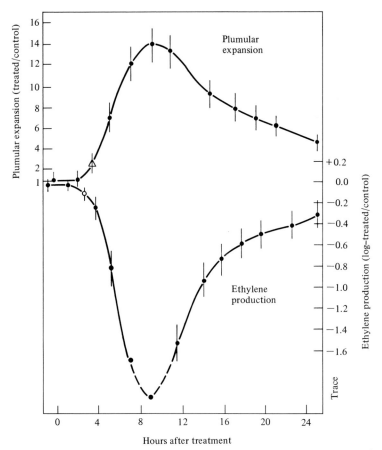

Figure 6-11. Effect of a 1-minute exposure to red light (40,000 ergs ·cm^{-1} total dose) on plumular expansion and hook opening and ethylene production in the upper part of the epicotyls in etiolated pea seedlings. (Redrawn, with permission, from Goeschl et al., 1967.)

Effects of Ethylene on Planes of Cell Expansion

Several of the effects of ethylene involve alteration of the normal planes of cell growth. Radial swelling or abnormal radial expansion of stems—such as is seen in the "triple response" of etiolated pea seedlings (Fig. 6-12) and probably also in leaf epinasty and in the herbicidal action of some auxin-type herbicides—and roots result from inhibited elongation and increased radial expansion of individual cells (Fig. 6-13). Sometimes "root hair-like" outgrowths and even adventitious roots are formed from

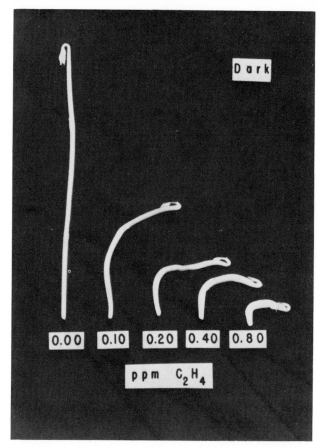

Figure 6-12. Effects of various concentrations of ethylene on the growth of etiolated pea seedlings, showing the portions of the seedlings which developed during the 48-hour treatment. (From Goeschl and Pratt, 1968.)

the cells. Radial expansion of cells which normally expand mainly by elongation is correlated with an alteration by ethylene of the orientation of the cellulose microfibrils in the cell wall. The newly deposited microfibrils at the inner surface of the wall are oriented in a more longitudinal plane than normal, restricting cell elongation but permitting radial expansion. An effect on auxin transport may also be involved in some of the abnormal growth caused by ethylene.

Other Effects of Ethylene

There are several other effects of ethylene which will be considered only briefly. One of these is of long historical interest and considerable eco-

Figure 6-13. Longitudinal sections of the cortical region of etiolated pea seedling stems. Left, control; right, ethylene-treated. (From Stewart et al., 1974, by permission; photos courtesy of M. Lieberman, 1978.)

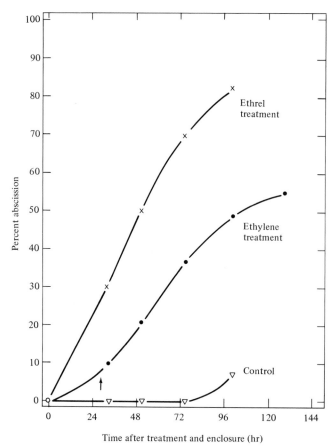

Figure 6-14. Abscission of leaves of cotton plants following treatment with ethylene and Ethrel (2-chloroethylphosphonic acid). (Redrawn, with permission, from Morgan, 1969.)

nomic importance. This is stimulation of flowering in mangos and bromeliads such as pineapple. For many years, the synthetic auxin naphthaleneacetic acid (NAA) was sprayed on pineapple fields to stimulate flowering. It turns out that it is NAA-stimulated ethylene which is the actual active hormone involved in the response.

Leaf abscission is another process which is stimulated by ethylene (Fig. 6-14), and many chemicals which are utilized as defoliants, including some auxin-type herbicides such as 2,4-D, act by inducing ethylene formation. As effective as exogenous ethylene is in causing leaf abscission, endogenous ethylene is not considered to be a major hormone involved in the natural process. Other hormones, such as auxin, gibberellins, cytokinins, and perhaps a yet unidentified "senescence factor," are more important.

Practical applications of ethylene on a commercial scale really got un-

$$Cl-CH_2-CH_2-\overset{\overset{O}{\|}}{\underset{\underset{O^-}{|}}{P}}-O^- + H_2O \ (or \ OH^-) \longrightarrow Cl-CH_2-CH_2-\overset{\overset{O^-}{|}}{P^+}-O^-$$

with

$$H-\overset{|}{\underset{\underset{H}{|}}{O}} \quad O^-$$

$$Cl^- + CH_2=CH_2 + H_2PO_4^- \ (or \ HPO_4^{2-})$$

Figure 6-15. 2-Chloroethylphosphonic acid, a synthetic growth regulator which decomposes spontaneously in aqueous solution and in plant tissues to yield ethylene and phosphoric acid. (Redrawn, with permission, from Varner, J. E. and D. T. Ho. 1976. Hormones. In: Bonner, J. and J. E. Varner, eds. *Plant Biochemistry. 3rd ed,* pp. 713–770. Copyright by Academic Press, Inc., New York.)

derway only as recently as about 1963, at which time a new water-soluble, slow ethylene-releasing growth regulator was introduced. Patented as "Ethrel" and given the trivial name "ethephon," the compound is 2-chloroethylphosphonic acid (Fig. 6-15). Ethephon decomposes spontaneously in aqueous solution and in tissues to yield ethylene. Within a few years after its introduction, this synthetic plant growth regulator was being utilized extensively on certain fruit and vegetable crops, including pineapple and cucumbers.

Ethylene Biosynthesis and Mechanism of Action

Ethylene evidently is much more restricted in its occurrence than the other kinds of plant hormones. It probably is ubiquitous among seed plants, but its distribution among other phyletic groups is limited. For certain, some bacteria and fungi produce ethylene, and ethylene production by certain fungal pathogens is directly involved in the host–pathogen interaction and the disease symptoms that develop in the host.

Ethylene is synthesized from methionine via S-adenosylmethionine as an intermediate, according to the pathway proposed recently from investigations with apple tissue (Fig. 6-16). Methionine is activated by ATP to give S-adenosylmethionine, which then is fragmented to give ethylene (from carbon atoms 3 and 4), formate (from carbon atom 2), carbon dioxide (from carbon atom 1), ammonium ion, and 5'-S-methyl-5'-thioadenosine. Subsequent work has shown that, in fact, S-adenosylmethionine first fragments to form 5'-S-methyl-5'-thioadenosine and a cyclic amino acid, 1-aminocyclopropane-1-carboxylic acid. The latter intermediate is then converted to ethylene, formate, CO_2, and ammonium ion. To complete the cycle, the nucleoside can donate its methylthio group to a 4-carbon acceptor (homoserine) to reform methionine.

As discussed in a previous section, the mechanism of action of ethyl-

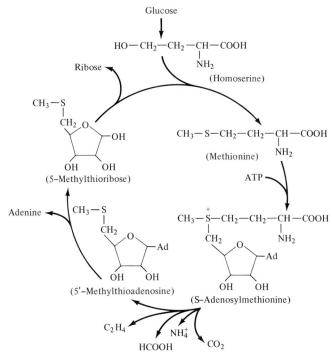

Figure 6-16. Proposed scheme of methionine metabolism in relation to ethylene biosynthesis in apple. (Redrawn, with permission, from Adams and Yang, 1977.)

ene is at present virtually unknown. Some of the effects of ethylene—such as promoting leaf abscission—involve de novo synthesis of particular enzymes (e.g., cellulase). However, it remains unknown as to how—whether directly or indirectly—ethylene stimulates enzyme synthesis. Moreover, the effects on membranes are incompletely resolved. Considerable evidence indicates that the hormone binds to a metal (perhaps copper) in whatever action it undergoes, but the details and metabolic consequences of such binding have yet to be elucidated.

References

Abeles, F. B. 1966a. Auxin stimulation of ethylene evolution. *Plant Physiol.* **41**: 585–588.

Abeles, F. B. 1966b. Effect of ethylene on auxin transport. *Plant Physiol.* **41**: 946–948.

Abeles, F. B. 1967a. Inhibition of flowering in *Xanthium pennsylvanicum* Walln. by ethylene. *Plant Physiol.* **42**: 608–609.

Abeles, F. B. 1967b. Mechanism of action of abscission accelerators. *Physiol. Plantarum* **20**: 442–454.

Abeles, F. B. 1972. Biosynthesis and mechanism of action of ethylene. *Ann. Rev. Plant Physiol.* **23**: 259–292.

Abeles, F. B. 1973. *Ethylene in Plant Biology.* Academic Press, New York.

Abeles, F. B. and H. E. Gahagan III. 1968. Abscission: The role of ethylene, ethylene analogues, carbon dioxide, and oxygen. *Plant Physiol.* **43**: 1255–1258.

Abeles, F. B. and R. E. Holm. 1966. Enhancement of RNA synthesis, protein synthesis, and abscission by ethylene. *Plant Physiol.* **41**: 1337–1342.

Abeles, F. B. and J. M. Ruth. 1972. Mechanisms of hormone action. Use of deuterated ethylene to measure isotopic exchange with plant material and the biological effects of deuterated ethylene. *Plant Physiol.* **49**: 669–671.

Adams, D. O. and S. F. Yang. 1977. Methionine metabolism in apple tissue. Implication of S-adenosylmethionine as an intermediate in the conversion of methionine to ethylene. *Plant Physiol.* **60**: 892–896.

Adams, D. O. and S. F. Yang. 1979. Ethylene biosynthesis: identification of 1-aminocyclopropane-1-carboxylic acid as an intermediate in the conversion of methionine to ethylene. *Proc. Natl. Acad. Sci. U.S.A.* **76**: 170–174.

Addicott, F. T. 1970. Plant hormones in the control of abscission. *Biol. Rev.* **45**: 485–524.

Andreae, W. A., M. A. Venis, F. Jursic, and T. Dumas. 1968. Does ethylene mediate root growth inhibition by indole-3-acetic acid? *Plant Physiol.* **43**: 1375–1379.

Apelbaum, A. and S. P. Burg. 1971. Altered cell microfibrillar orientation in ethylene-treated *Pisum sativum* stems. *Plant Physiol.* **48**: 648–652.

Beyer, E. M., Jr. and P. W. Morgan. 1970. Effect of ethylene on the uptake, distribution, and metabolism of indoleacetic acid-1-^{14}C and -2^{14}C and naphthaleneacetic acid-1-^{14}C. *Plant Physiol.* **46**: 157–162.

Biale, J. B. 1950. Postharvest physiology and biochemistry of fruits. *Ann. Rev. Plant Physiol.* **1**: 183–206.

Burg, S. P. 1962. The physiology of ethylene formation. *Ann. Rev. Plant Physiol.* **13**: 265–302.

Burg, S. P. 1968. Ethylene, plant senescence, and abscission. *Plant Physiol.* **43**: 1503–1511.

Burg, S. P. 1973. Ethylene in plant growth. *Proc. Natl. Acad. Sci. U.S.A.* **70**: 591–597.

Burg, S. P. and E. A. Burg. 1962. Role of ethylene in fruit ripening. *Plant Physiol.* **37**: 179–189.

Burg, S. P. and E. A. Burg. 1964. Biosynthesis of ethylene. *Nature (London)* **203**: 869–870.

Burg, S. P. and E. A. Burg. 1965a. Gas exchange in fruits. *Physiol. Plantarum* **18**: 870–884.

Burg. S. P. and E. A. Burg. 1965b. Ethylene action and the ripening of fruits. *Science* **148**: 1190–1196.

Burg, S. P. and E. A. Burg. 1965c. Relationship between ethylene production and ripening in bananas. *Bot. Gaz.* **126**: 200–204.

Burg, S. P. and E. A. Burg. 1966a. Auxin-induced ethylene formation: its relation to flowering in the pineapple. *Science* **152**: 1269.

Burg, S. P. and E. A. Burg. 1966b. The interaction between auxin and ethylene and its role in plant growth. *Proc. Natl. Acad. Sci. U.S.A.* **55**: 262–269.

Burg, S. P. and E. A. Burg. 1967a. Molecular requirements for the biological activity of ethylene. *Plant Physiol.* **42**: 144–152.

Burg, S. P. and E. A. Burg. 1967b. Inhibition of polar auxin transport by ethylene. *Plant Physiol.* **42**: 1224–1228.

Burg, S. P. and E. A. Burg. 1968. Ethylene formation in pea seedlings; its relation to the inhibition of bud growth caused by indole-3-acetic acid. *Plant Physiol.* **43**: 1069–1074.

Burg, S. P. and C. O. Clagett. 1967. Conversion of methionine to ethylene in vegetative tissue and fruits. *Biochem. Biophys. Res. Commun.* **27**: 125–130.

Chadwick, A. V. and S. P. Burg. 1967. An explanation of the inhibition of root growth caused by indole-3-acetic acid. *Plant Physiol.* **42**: 415–420.

Chadwick, A. V. and S. P. Burg. 1970. Regulation of root growth by auxin-ethylene interaction. *Plant Physiol.* **45**: 192–200.

Crocker, W., A. E. Hitchcock, and P. W. Zimmerman. 1935. Similarities in the effects of ethylene and the plant auxins. *Contrib. Boyce Thompson Inst.* **7**: 231–248.

Crocker, W., P. W. Zimmerman, and A. E. Hitchcock. 1932. Ethylene-induced epinasty of leaves and the relation of gravity to it. *Contrib. Boyce Thompson Inst.* **4**: 177–218.

Dilley, D. R. 1969. Hormonal control of fruit ripening. *HortScience* **4**: 111–114.

Eisinger, W. R. and S. P. Burg. 1972. Ethylene-induced pea internode swelling. Its relation to ribonucleic acid metabolism, wall protein synthesis, and cell wall structure. *Plant Physiol.* **50**: 510–517.

Frenkel, C., I. Klein, and D. R. Dilley. 1968. Protein synthesis in relation to ripening of pome fruits. *Plant Physiol.* **43**: 1146–1153.

Galston, A. W. and P. Kaur. 1961. Comparative studies on the growth and light sensitivity of green and etiolated pea stem sections. In: McElroy, W. D. and B. Glass, eds. *Light and Life.* Johns Hopkins Press, Baltimore. Pp. 687–705.

Goeschl, J. D., H. K. Pratt, and B. A. Bonner. 1967. An effect of light on the production of ethylene and the growth of the plumular portion of etiolated pea seedlings. *Plant Physiol.* **42**: 1077–1080.

Goeschl, J. D. and H. K. Pratt. 1968. Regulatory roles of ethylene in the etiolated growth habit of *Pisum sativum.* In: Wightman, F. and G. Setterfield, eds. *Biochemistry and Physiology of Plant Growth Substances.* Runge Press, Ottawa. Pp. 1229–1242.

Hansen, E. 1966. Postharvest physiology of fruits. *Ann. Rev. Plant Physiol.* **17**: 459–477.

Hansen, E. and E. Hartman. 1937. Effect of ethylene and metabolic gases upon respiration and ripening of pears before and after cold storage. *Plant Physiol.* **12**: 441–454.

Jackson, M. B. and D. J. Campbell. 1976. Waterlogging and petiole epinasty in tomato: the role of ethylene and low oxygen. *New Phytol.* **76**: 21–29.

Kang, B. G., W. Newcomb, and S. P. Burg. 1971. Mechanism of auxin-induced ethylene production. *Plant Physiol.* **47**: 504–509.

Kang, B. G., C. S. Yokum, S. P. Burg, and P. M. Ray. 1967. Ethylene and carbon dioxide: Mediation of hypocotyl hook-opening response. *Science* **156**: 958–959.

Kawase, M. 1976. Ethylene accumulation in flooded plants. *Physiol. Plantarum* **36**: 236–241.

Kende, H. and A. D. Hanson. 1976. Relationship between ethylene evolution and senescence in morning-glory flower tissue. *Plant Physiol.* **57**: 523–527.

Kidd, F. and C. West. 1933. The influence of the composition of the atmosphere upon the incidence of the climacteric in apples. *Rep. Food Invest. Board.* 51–57.

Leopold, A. C. and P. E. Kriedemann. 1975. *Plant Growth and Development*. 2nd ed. McGraw-Hill Book Company, New York.

Lieberman, M. 1975. Biosynthesis and regulatory control of ethylene in fruit ripening. A review. *Physiol. Vég.* **13**: 489–499.

Lieberman, M., A. Kunishi, L. W. Mapson, and D. A. Wardale. 1966. Stimulation of ethylene production in apple tissue slices by methionine. *Plant Physiol.* **41**: 376–382.

Mapson, L. W. 1970. Biosynthesis of ethylene and the ripening of fruit. *Endeavour* **29**: 29–33.

McGlasson, W. B. 1970. The ethylene factor. In: Hulme, A. C., ed. *The Biochemistry of Fruits and Their Products*. Vol. 1. Academic Press, New York. Pp. 475–519.

Morgan, P. W. 1969. Stimulation of ethylene evolution and abscission in cotton by 2-chloroethanephosphonic acid. *Plant Physiol.* **44**: 337–341.

Morgan, P. W. and H. W. Gausman. 1966. Effects of ethylene on auxin transport. *Plant Physiol.* **41**: 45–52.

Phillips, I. D. J. 1971. *Introduction to the Biochemistry and Physiology of Plant Growth Hormones*. McGraw-Hill Book Company, New York.

Pratt, H. K. and J. D. Goeschl. 1969. Physiological roles of ethylene in plants. *Ann. Rev. Plant Physiol.* **20**: 541–584.

Ray, P. M. 1963. *The Living Plant*. Holt, Rinehart & Winston, New York.

Rubinstein, B. and F. B. Abeles. 1965. Relationship between ethylene evolution and leaf abscission. *Bot. Gaz.* **126**: 255–259.

Stewart, R. N., M. Lieberman, and A. T. Kunishi. 1974. Effects of ethylene and gibberellic acid on cellular growth and development in apical and subapical regions of etiolated pea seedlings. *Plant Physiol.* **54**: 1–5.

Varner, J. E. and D. T. Ho. 1976. Hormones. In: Bonner, J. and J. E. Varner, eds. *Plant Biochemistry*. 3rd ed. Academic Press, New York. Pp. 713–770.

Von Abrams, G. J. and H. K. Pratt. 1967. Effect of ethylene on the permeability of excised cantaloupe tissue. *Plant Physiol.* **42**: 299–301.

Wang, C. Y. and W. M. Mellenthin. 1972. Internal ethylene levels during ripening and climacteric in Anjou pears. *Plant Physiol.* **50**: 311–312.

Zimmerman, P. W. and F. Wilcoxon. 1935. Several chemical growth substances which cause initiation of roots and other responses in plants. *Contrib. Boyce Thompson Inst.* **7**: 209–229.

Chapter 7

Phytochrome

Introduction

Phytochrome very definitely is not a hormone. It is a pigment. However, no book on plant hormones could be regarded as complete without detailed consideration of phytochrome. For phytochrome is involved in the regulation of growth and development literally throughout the ontogeny of seed plants, from the germination of some kinds of seeds to floral initiation in many species of angiosperms. There are multiple important interactions of the phytochrome system with hormone metabolism and hormone action and complex interactions as well between the phytochrome system and endogenous circadian rhythms.

History of Discovery and Modern Description

Phytochrome is a water-soluble chromoprotein which exists in two definitive, photointerconvertible forms—P_r, which absorbs light maximally in the red region ($\lambda_{max} = 660$ nm), and P_{fr}, which absorbs maximally in the near far-red ($\lambda_{max} = 730$ nm)—and which serves as the photoreceptor in all the "red, far-red reversible photoreactions" of plants

The definitive discovery of phytochrome in 1959 by H. A. Borthwick, W. L. Butler, S. B. Hendricks, K. H. Norris, and H. W. Siegelman at U.S.D.A. laboratories at Beltsville, Maryland constitutes one of the most significant achievements in plant physiology. And how fitting it is that phytochrome ultimately was discovered at the same laboratories where W. W. Garner and H. A. Allard discovered photoperiodism some 39 years

Table 7-1. Sample Class Data Showing the Effects of Red and Far-red Irradiation on the Germination of Grand Rapids Lettuce Seeds[a]

Treatment	Percentage germination
Dark control	5
Red	93
Red–far-red	11
Red–far-red–red	89

[a] Seeds were hydrated for 11 hours in darkness at room temperature. Working under a blue safelight in a darkroom, 50 seeds each were placed on moist filter paper in four Petri dishes. One group of seeds then was placed in darkness, and the others were irradiated using incandescent light passed through red and far-red glass interference filters. Exposures to red light were 4 minutes, and to far-red light, 8 minutes. All groups were returned to darkness after irradiation, and percentages germination were determined 48 hours after irradiation.

earlier. The story of how phytochrome was discovered should be appreciated by all students of plant biology.

A very fundamental advance in plant photobiology, which dates from circa 1935, was the discovery that various plant responses to light are elicited maximally by red light, and that certain of these actions of red light are reversible by far-red light. For example, it has been known since the early reports by Flint and McAlister (1935, 1937) that the germination of some kinds of seeds which require light (e.g., lettuce, *Latuca sativa* var. Grand Rapids) is promoted by red light and suppressed by far-red light. Detailed action spectra for the germination response of light-sensitive lettuce seeds showed maximum promotion near 660 nm and maximum suppression near 730 nm. Additionally, it was found that the effects of red light were experimentally reversible by immediately following a red light treatment with far-red irradiation. By repeatedly alternating brief red and far-red treatments, it was seen that the light applied last determined whether the seeds germinated or not (Table 7-1). Within approximately a decade, it was discovered that interruption of a long dark period inhibited or prevented flowering of short-day plants and promoted flowering of long-day plants, and that the action spectrum for this "night-break" or "night interruption" phenomenon was identical to that for promotion of light-sensitive seed germination. Moreover, red, far-red reversibility also obtained in the "night-break" phenomenon.

The simplest interpretation of these photoresponses—one photoperiodic, one nonphotoperiodic—was that a single photoreceptor pigment was involved which could exist in two photointerconvertible forms:

$$P_r \xrightleftharpoons[\text{far-red light}]{\text{red light}} P_{fr}$$

Since the energy requirements for the photoreactions were relatively quite low (as compared to photosynthesis, for example), it seemed certain that the effective wavelengths brought about changes only in molecular configuration of the pigment per se and did not involve transfer of energy to another system. Furthermore, since red light potentiates responses and far-red nullifies the action of red light, it was hypothesized (correctly, as it turns out) that only one form (the far-red light absorbing form, P_{fr}) of the pigment should be active, the other inactive. Thus, it came to be understood that the response of a plant to red light (that is, in the red, far-red reversible responses) results from the action of P_{fr} thereby generated. Response to far-red results from the absence of action of P_{fr}.

Attempts to demonstrate directly the pigment in plants and to extract it from plant tissues began in earnest in the mid-1950s. The problem was understandably difficult for various reasons: (1) the pigment occurs in very low concentrations in plant tissues (e.g., 10^{-8} to 10^{-7} M); (2) passing a beam of red or far-red light through a sample of the pigment (in vivo or in vitro) to detect or measure one form of the pigment converts it to the other form; (3) detection and measurement in vivo are complicated by the fact that plant tissues are optically very dense, highly light-scattering materials; (4) the presence of chlorophyll in green tissues and the conversion of protochlorophyll to chlorophyll upon irradiation of etiolated tissues pose obvious problems also; and (5) the probability that the pigment was a chromoprotein made efforts to reintroduce extracts into living tissues seem unpromising, so bioassay was excluded. Thus, a special spectrophotometer capable of measuring very slight differences in absorption in very dense materials had to be built so advantage could be taken of the reversible absorbance by the pigment of red and far-red light. The instrument that was developed was a dual-wavelength difference spectrophotometer, which permitted a direct measurement of the ΔOD between 660 and 730 nm.

Finally, in 1959, the definitive discovery of the pigment, then named "phytochrome," was made by W. L. Butler, K. H. Norris, H. W. Siegelman, and S. B. Hendricks at Beltsville. They successfully detected phytochrome in vivo in the cotyledons of etiolated turnip seedlings and in shoots of etiolated corn seedlings. They also extracted and partially purified the pigment from etiolated corn seedlings. The first rigorous efforts to purify phytochrome extensively were by Siegelman and Firer (1964), who achieved about 60-fold purification by conventional methods of protein purification. With methods used subsequently, purifications of 750-fold or better are routinely achieved. The pigment can easily be seen on columns during purification and changes color upon irradiation with red and far-red light. P_r is blue and P_{fr} is blue-green. Absorption spectra for partially

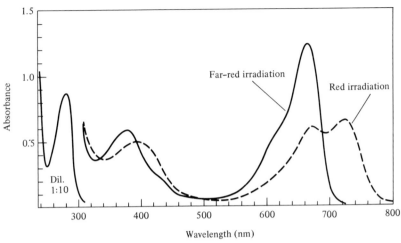

Figure 7-1. Absorption spectra of a solution of oat phytochrome following irradiation with red and far-red light. The spectrum observed after far-red irradiation is the absorption spectrum for P_r; that observed after red irradiation is the absorption spectrum for P_{fr}. The spectrum at wavelengths shorter than 300 nm was measured on a solution diluted 10-fold. (Reproduced, with permission, from the *Annual Review of Plant Physiology, Volume 16.* © 1965 by Annual Reviews Inc. From Siegelman and Butler, 1965.)

purified P_r and P_{fr}—measured after forcing irradiation with far-red and red light, respectively—show absorption peaks for P_r at about 660 nm in the red, and for P_{fr} at about 730 nm (Fig. 7-1). Notably also, both P_r and P_{fr} show significant absorption in the blue region, the absorption maxima being at about 370 and 400 nm, respectively. The action spectra for the phototransformations of P_r and P_{fr} are as predicted (Fig. 7-2). With narrow-band far-red light, P_{fr} can be almost completely transformed to P_r. However, because of the overlapping of the absorption spectra in the region of 600 to 700 nm, the most effective attainable band of red light transforms only about 80% of P_r to P_{fr}.

Basically four major transformation reactions have been described for phytochrome (Fig. 7-3). Reactions I and II, described previously, both are first-order, photochemical reactions. Reaction IV, a zero-order destruction or decay reaction, seems to be ubiquitous in all tissues containing phytochrome. It results in a decrease in total photoreversible phytochrome, probably in fact degradation of the protein moiety. Dark reversion of P_{fr} to P_r, reaction III, has been demonstrated in various dicots but as yet has not been demonstrated in monocots.

The quantum efficiencies for the two photoreactions are such that reaction I is energetically over twice as efficient as reaction II (Fig. 7-2). The action of blue light is very much less effective than red and far-red light in reactions I and II. Red light has been reported to be 100 times more effec-

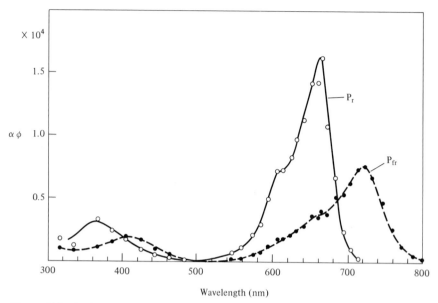

Wavelength (nm)

Figure 7-2. Action spectra of photochemical transformations of P_r and P_{fr}. The extinction coefficient, α, is in liter mol^{-1} cm^{-1}, and the quantum yield, ϕ, is in moles Einstein^{-1}. (Reproduced, with permission, from the *Annual Review of Plant Physiology, Volume 16*. © 1965 by Annual Reviews Inc. From Siegelman and Butler, 1965.)

tive than blue, in vivo in converting P_r, and far-red light is 25 times more efficient than blue in converting P_{fr}. In vitro, blue light is relatively more effective than in vivo but still markedly less effective than red and far-red light.

As noted previously, P_{fr} is the active form of phytochrome, and P_r is metabolically inert. Thus, in phytochrome-mediated photoresponses, a response to irradiation with red light results from the action of P_{fr} which is thereby generated or maintained. Conversely, an effect of far-red light results from the absence of action of P_{fr}—not from any action of P_r. Notably, white light or sunlight has a net red-light effect, even though such radiation of course contains wavelengths absorbed by both forms of phytochrome. The main reason for this is the greater quantum efficiency of reaction I than reaction II.

Phytochrome is a water-soluble chromoprotein of imprecisely determined molecular weight, perhaps a monomer molecular weight of approximately 120,000. The probable chromophore is a bilitriene or open-chain tetrapyrrole which is closely similar to but distinct from the chromophore of c-phycocyanin and allophycocyanin (Fig. 7-4). Available evidence suggests that there probably is one chromophore per molecule of phytochrome. The chromophore is attached to the protein by an ester bond

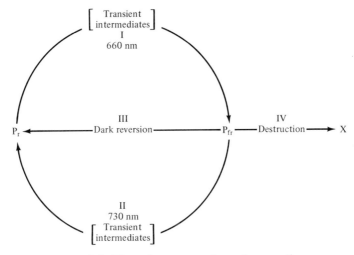

Figure 7-3. Phytochrome transformation reactions.

involving ring III and an unidentified group in the protein and probably by another bond involving ring I. Upon phototransformation of P_r to P_{fr}, a proton migration from ring I to its ethylidene group occurs, causing a shift in a double bond, and the loss of another hydrogen atom. Subtle changes evidently occur also in the protein, which probably accounts for P_{fr} being active and P_r being inactive.

Figure 7-4. Proposed structure for the chromophore of phytochrome, its linkage to the protein, and photo-transformation mechanism. [Redrawn, with permission, from Rüdiger, W. 1972. Chemistry of phytochrome chromophore. In: Mitrakos, K. and W. Shropshire, Jr., eds. *Phytochrome*. Pp. 129–141. Copyright by Academic Press, Inc. (London) Ltd.]

Occurrence, Distribution, and
Intracellular Localization

Phytochrome has been positively identified in numerous species of angiosperms, gymnosperms, green algae, liverworts, and plants of other phyletic groups. It seems safe to conclude, in fact, that the pigment occurs in all green plants. Quite probably also, it is physiologically active in all chlorophyllous plants.

The most intensive investigation of phytochrome distribution within seedlings was conducted by W. R. Briggs and H. W. Siegelman (1965). Using in vivo spectroscopy, they determined distribution in etiolated barley, corn, oat, pea, bean, and sunflower seedlings (Fig. 7-5). In general, the highest concentrations of phytochrome were found in tissues which were either meristematic or which had recently been meristematic. Regions with the highest levels of detectable phytochrome also had the highest phytochrome–protein ratios, suggesting that these regions might be active sites of phytochrome synthesis. The distribution of phytochrome was compared with that of other physiological systems, and found generally to parallel auxin distribution, phototropic sensitivity, and red light sensitivity.

As regards intracellular localization of phytochrome, S. B. Hendricks and H. A. Borthwick first suggested, on the basis of some very rapid phytochrome-mediated responses which involve alterations in membrane permeability, that phytochrome might exist in association with membranes. Available evidence continues to support the notion that some but not all phytochrome is associated with some part of the membrane system of cells, including the plasmalemma and membranes of etio-

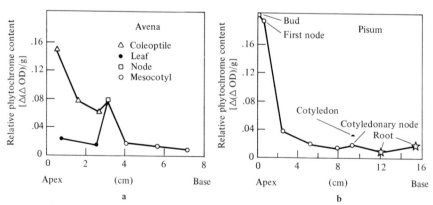

Figure 7-5. Distribution of phytochrome in dark-grown oat shoots, **a**, and dark-grown pea seedlings, **b**, per gram fresh weight. (Redrawn, with permission, from Briggs and Siegelman, 1965.)

plasts and mature chloroplasts. There is considerable evidence also that P_{fr} becomes rapidly attached to some component of the membrane system after it is formed by red light.

In the tissues of most plants, P_{fr} appears to be labile, as previously noted (Fig. 7-3), undergoing a nonphotochemical transformation known as destruction. Destruction evidently occurs in both light and darkness, but it generally occurs more rapidly in light because of the constant generation of P_{fr} from P_r. Even though new phytochrome is synthesized, as P_r, as growth proceeds, it still is important to determine whether and how effective levels of phytochrome are maintained. Certainly a large portion of the phytochrome present in a dark-grown seedling is lost upon the first exposure to light. D. T. Clarkson and W. S. Hillman (1967) reported on some experiments with pea seedlings which helped to clarify this point. They found that upon irradiation of decapitated, etiolated pea seedling shoots the concentration of total photoreversible phytochrome (P_r and P_{fr}) decreased to about 10% of the original level (Fig. 7-6). At that point a

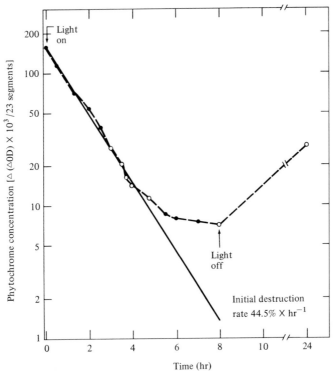

Figure 7-6. In vivo measurements of phytochrome concentration during the course of continuous illumination of decapitated, etiolated pea plants by red light, and after a subsequent dark period. (Redrawn, with permission, from Clarkson and Hillman, 1968.)

steady-state level was maintained in the light as synthesis of new P_r counteracted P_{fr} destruction. When the light was turned off, the concentration of phytochrome began to increase as the rate of synthesis of new P_r exceeded the rate of destruction of P_{fr}.

Induction–Reversion versus High Irradiance Responses

It is well to pause here to note that there are in fact two distinct types of phenomena which have been attributed to phytochrome—the so-called "induction–reversion" responses, with which this chapter is chiefly concerned, and the "high irradiance" responses. The former are the more familiar responses which are "light-triggered," which is to say that they are responses which are induced by a brief irradiation with relatively low intensity red light and reversible by immediate subsequent irradiation with far-red light. Light is viewed as merely "triggering" the response and as having no direct role in the inductive action of P_{fr}. Under conditions of nonsaturating light doses, these responses obey the law of reciprocity, according to which, for any given wavelength, intensity multiplied by time of irradiation is a constant. These are the classical phytochrome-mediated responses, and the photoreversibility of these responses is the generally accepted criterion, in fact, for establishing phytochrome involvement in a plant photoresponse.

High irradiance responses, in contrast, are "light driven" photoresponses in the case of which continuous irradiation—with red, far-red, or blue light—at relatively quite high intensity is necessary to sustain the response. Light is viewed as having a direct role in the inductive action of P_{fr}. They are effects which are observed only where a photostationary state of $P_r : P_{fr}$ is established and maintained over relatively long periods. Reciprocity does not apply, and red, far-red photoreversibility is not always demonstrable. If irradiation ceases, the response simply reverts to the control condition, without the necessity of a final antagonistic irradiation. Of the high irradiance responses known, that of photoinhibition of lettuce hypocotyl elongation probably is the most studied. The action spectrum for this response shows a single sharp, symmetrical peak at approximately 720 nm. Phytochrome may play a fundamentally different role in mediating the light effect in the high irradiance responses than in the "induction–reversion" responses to which most of this chapter is devoted. Moreover, where both blue and far-red action peaks are found, the blue peak must, in fact, be attributable to some pigment other than phytochrome.

Non-Phytochrome Mediated Photoresponses to Blue Light

There is, in fact, another entirely separate photoreceptor system in lower and higher plants which many investigators consider to be equally as important as the phytochrome system. This "blue light photoreceptor" (Briggs, 1976) absorbs light in the long-wavelength ultraviolet and blue regions, and there is none of the photoreversibility that is characteristic of many phytochrome-mediated responses. Exact wavelengths vary somewhat from organism to organism, but the responses mediated by the blue light photoreceptor are always identical in the following features: a single broad peak near 370 nm, a shoulder near 420 nm, an action maximum between 445 and 460 nm, and another peak or shoulder between 470 and 490 nm. Such action spectra are typical, for examples, for phototropism of *Avena* coleoptiles (first and second positive curvatures) and *Phycomyces* sporangiophores and for carotenoid synthesis in certain fungi. The blue light photoreceptor itself probably is a flavoprotein which is photoreduced by light, thereupon reduces a b-type cytochrome, and is reoxidized in following darkness.

Introduction to Mechanism of Phytochrome Action

It has been emphasized already that phytochrome is the photoreceptor for all of the red, far-red reversible photoreactions of plants. The many photoreactions or processes involving phytochrome are of three major types, namely: (1) photoperiodic, photomorphogenetic responses, such as floral initiation in photoperiodically sensitive species, onset of winter bud dormancy, tuber formation, and many others; (2) nonphotoperiodic, photomorphogenetic responses, such as greening of etiolated seedlings, germination of some kinds of seeds, elongation of fern rhizoids, and others; and (3) nonmorphogenetic photoresponses, such as synthesis of anthocyanins, induction of various chloroplastic enzymes, and others.

Some concepts of phytochrome action envisage a single mechanism of action of P_{fr} in all the many photoresponses for which phytochrome is the photoreceptor, and other concepts encompass two or more mechanisms of action. Whichever concept is favored, on the basis of information presently available, it is expedient to consider the action of phytochrome in the generally longer term and inductive photoperiodic reactions separately from the generally more rapid, noninductive photoreactions. This is the approach used in the following sections. We shall first consider the role of phytochrome in several of the noninductive photoresponses. Next, the evidence that P_{fr} stimulates GA biosynthesis and efflux from plastids will be presented. Finally, we will turn to a discussion of the role of phytochrome in photoperiodism.

Phytochrome Action in Nonphotoperiodic Photoresponses

Effects on Membrane Permeability

An idea that originated with H. A. Borthwick and his group at Beltsville circa 1966, on the basis of some investigations of nyctinastic leaf movements, is that phytochrome (specifically the active form, P_{fr}) might act by regulating membrane permeability. Subsequent investigations by many scientists have substantiated repeatedly the fact that this is indeed a function of phytochrome in many photoresponses.

Fondeville et al. (1966) investigated "sleep movements" (nyctinasty) in the leaves of *Mimosa pudica,* a legume having doubly pinnately compound leaves. In this plant and other related legumes (e.g., *Albizzia julibrissin,* Fig. 7-7) the pinnules of the doubly compound leaves are open and separate in the light but "fold together" at night. The leaf movements result directly from differential turgor changes in structures called pulvini. In a doubly pinnately compound leaf, there is a primary pulvinus at the base of the rachis, a secondary pulvinus at the base of each rachilla, and a tertiary pulvinus (or pulvinule) at the base of each pinnule (Fig. 7-7). Fondeville et al. used excised pinnae, each consisting of approximately 12 paired pinnules, and observed opening and closing of the paired pinnules on the axis of the pinna.

Action spectra were determined for the potentiation of the closing movements and its reversal. If the plants were irradiated with red light immediately before a dark period, the leaves began to close within 5 minutes and were fully closed in 30 minutes (Fig. 7-8). But if the plants were irradiated with far-red light immediately prior to a dark period, the leaves remained open for several hours in darkness. The potentiated control of closing movements could be repeatedly established and reversed by alternating irradiations with red and far-red light, respectively.

Nyctinastic leaf movements are caused by relative changes in turgor pressure (and also cell size) on opposite sides of the pulvinus (or pulvinule). In darkness there is a movement of water out of the cells in the upper side of the pulvinus into cells in the lower side, causing a folding or closing movement of the leaflets. In the light, this process is reversed. The transfer of water is caused by an apparent alteration of membrane perme-

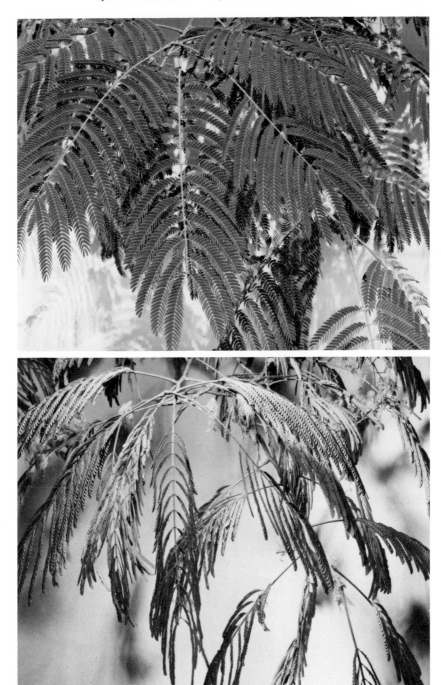

Figure 7-7. *Albizzia julibrissin* leaves in normal daytime position (top) and in typical sleep (night) position (bottom).

Figure 7-8. Pinnae of *Mimosa pudica* 30 minutes after placing in darkness following high-intensity light terminated with a succession of 2-minute exposures to red (r) and far-red (fr) irradiation. The pinnae remained open if exposure to far-red was last (top row) and closed if red irradiation was last (bottom row). (With permission, from Fondeville et al., 1966; photo courtesy of James D. Anderson, 1978.)

ability and the transfer of K^+ and Cl^- ions (perhaps also other electrolytes). The cells comprising pulvini contain extraordinarily high concentrations of K^+, on the order of $0.5\ M$. Leaflet closure is associated with a loss of K^+ and Cl^- from the upper cells and a concomitant loss of water. The lower cells evidently do not gain the ions lost from the upper cells, but the lower cells, because of their negative water potential, do gain water lost from the upper cells. The role of phytochrome in all of this is either to cause a differential transient increase in the permeability of the plasma membranes of the upper cells, or perhaps to activate a membrane-bound ATPase.

An interesting and physiologically important effect of phytochrome on phototaxis of chloroplasts in the filamentous green alga *Mougeotia* was reported by W. Haupt in 1963. Using microbeams of monochromatic light, Haupt observed that, when focused upon or even very near the plate-like chloroplast in a single cell, the chloroplast would orient perpendicular to a microbeam of low-intensity red light and parallel to a beam of far-red light (Fig. 7-9). The effects of red and far-red light were fully reversible when alternating irradiations were used. Moreover, the change in plastid orientation was clearly evident in less than 10 minutes and was

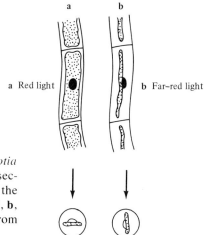

Figure 7-9. Diagram of the alga *Mougeotia* (longitudinal section above, transverse section below), illustrating phototaxis of the plate-like chloroplast in red, **a**, and far-red, **b**, light. (Redrawn, with permission, from Haupt, 1963.)

complete in 30 minutes. This phenomenon has yet to be elucidated, and there may be no effect of phytochrome on membrane permeability involved in this case. However, it is another example of a rapid kind of phytochrome-mediated response which evidently does not involve any effect of phytochrome on RNA or protein synthesis.

Further evidence for phytochrome being associated with cell membranes and having an effect on membrane properties has come from investigations of the so-called "Tanada effect." The empirical observations involved are as follows: Excised root tips from a plant such as mung bean (*Phaseolus aureus*) are suspended in solution (containing IAA, ATP, ascorbic acid, K^+, Mg^{2+}, Mn^{2+}, Ca^{2+}, and Cl^-) in a beaker in which is situated vertically a glass plate, the surface of which is made negative by washing with phosphate ions. When irradiated with red light, it is observed that the root tips adhere to the glass plate when the suspension is stirred. When irradiated with far-red light, the root tips release from the negatively charged surface (Fig. 7-10). By direct measurement it was confirmed that in red light the root tips developed a positive bioelectric potential at the tip, and in far-red light they developed a negative bioelectric potential. It was concluded that phototransformation of phytochrome changes the permeability characteristics of the cell membranes, with which the phytochrome probably is associated, resulting in a localized electrochemical gradient manifested as a bioelectric potential. Perhaps P_{fr} somehow affects a plasma membrane-bound ATPase, since K^+ and H^+ transport across the membrane is facilitated by such an enzyme.

Differential Gene Activation

The responses described in the preceding section are all relatively rapid phenomena which evidently involve an effect of phytochrome on properties of cell membranes, with which the photoreceptor probably is inti-

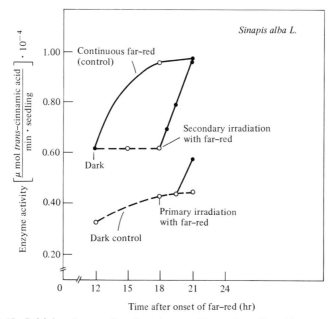

Figure 7-13. Initial and secondary lag-phases of far-red mediated increase of phenylalanine ammonia lyase (PAL) activity in the white mustard seedling. To determine a secondary lag-phase the seedlings were irradiated for 12 hours with standard far-red, placed in darkness for 6 hours, and then re-irradiated with the same standard far-red. (Redrawn, with permission, from Mohr et al., 1968.)

That is, PAL activity increased immediately after the onset of the second irradiation (Fig. 7-13). Mohr and the Freiburg group have interpreted their results as indicating that P_{fr} evokes de novo synthesis of PAL. However, evidence reported by Attridge and Johnson (1976) suggested, on the contrary, that synthesis of PAL was not involved but rather activation of pre-existing (but inactive) PAL in the tissues.

The evidence for stimulation of de novo synthesis of some other enzymes by P_{fr}—but at the translational, not the transcriptional level—is less controversial than for PAL. The principal evidence is that red light (specifically P_{fr} thereby generated) causes conversion of monoribosomes to polyribosomes in certain tissues prior to photomorphogenetic responses, e.g., light-sensitive lettuce seeds, etiolated leaves, and pine seeds. There evidently is some sort of activating effect even on the monoribosomes, since monoribosomes isolated from corn leaves preirradiated with red light were found to be more effective than monoribosomes from nonirradiated tissues in protein synthesis in vitro. In the case of etiolated bean leaves P_{fr} enhanced the conversion of monoribosomes to polyribosomes, from both cytoplasm and chloroplasts, even when mRNA synthesis was inhibited by cordycepin.

Thus, by way of summary, it can be concluded tentatively that P_{fr} acts in some cases by activation of specific enzymes, in others by evoking enzyme synthesis, and in still others by altering properties of membranes. Alteration of permeability properties quite possibly is itself a consequence of activation of a specific membrane-bound enzyme (ATPase).

Effect on GA Biosynthesis and Efflux from Plastids

Compartmentation of GA biosynthesis in plastids and stimulation of GA biosynthesis by red light were discussed in Chapter 3. It is appropriate to consider now the extensive data showing an important interaction of the phytochrome system with the GA economy of plants.

Certain effects of phytochrome on GA metabolism in plastids have been studied intensively. Stoddart (1968) reported that extracts of chloroplast fractions of leaves of *Brassica oleracea* and *Hordeum vulgare* had biological activity resembling that of GAs. Subsequently he and others presented evidence that at least some GA biosynthesis is compartmentalized in chloroplasts and other plastids. There have been numerous reports since 1968 also indicating that light alters GA metabolism, including apparent effects on biosynthesis, interconversions, apparent release from membrane-bound forms, and phytochrome-mediated efflux from plastids. Involvement of plastids in the rapid effect of red light on GA metabolism (see Chapter 3) was based initially on the response of etioplasts in vitro and in situ to red and far-red light, suggesting the presence of phytochrome within or on the plastid envelope. Evans and Smith (1976a,b) presented spectrophotometric evidence for the presence of phytochrome in the envelope membrane of barley etioplasts. Earlier, Cooke and Saunders (1975) had demonstrated a phytochrome dependent increase in extractable GA-like activity in plastid preparations from etiolated wheat leaves, and it was reported a short time later that phytochrome and GA-like substances both were associated with etioplast envelopes (Cooke and Kendrick, 1976). Most recently, GA_4 and GA_9 were shown to be tenaciously associated with chloroplast membranes, since the GAs were released by detergent treatment but not by sonication (Browning and Saunders, 1977).

Ellis and Hartley (1971) suggested that there must be a specific mechanism for transporting chloroplast proteins made on cytoplasmic ribosomes and other substances across the outer membrane of plastids. Recent work suggests that the hypothetical transmembrane transport mechanism involves phytochrome. Many workers have reported a direct relation between irradiation with red light and the level of extractable GA-like activity or actual GAs from leaves of different species. This effect has been shown to be phytochrome dependent. Evans and Smith (1976a,b) demonstrated the presence of phytochrome within the etioplast and its relationship to the increase in extractable GA upon illumination. They concluded that phytochrome would cause the transport of GA across the etio-plast envelope. And, indeed, the transport of different compounds, including GAs, across plastid envelopes, has been reported by Wellburn and Hampp (1976) and the regulation of this process by phytochrome has been discussed (Hampp and Schmidt, 1977; Schmidt and Hampp, 1977). From the available data, we can conclude that some of the effects of red light (of P_{fr}) are specifically correlated with increases in free GAs available for growth regulation.

Phytochrome and Photoperiodic (Flowering) Responses

Introduction

Experiments were conducted in the 1940s on interrupting dark periods by flashes or longer exposures to light to ascertain effects on the flowering of photoperiodically sensitive species. This "night break" or "night inter-ruption" phenomenon is manifested as a nullification of the promotive ef-fect of an otherwise inductive dark period for certain plants (the "short-day plants") and nullification of the inhibitory effect of the dark period for other plants (the "long-day plants"). By the early 1950s—soon after the action spectra for lettuce seed germination were worked out—it was ob-served that red light is most effective in the night interruption phenome-non, and that the effect of red can be nullified by immediate subsequent irradiation with far-red light. Indeed, it was the nature of the light-sensi-tive seed germination response and of the night interruption phenomenon on which the theoretical existence of a pigment system such as phy-tochrome first was postulated. Promptly upon the definitive discovery of phytochrome in 1959, it could be confidently concluded that phytochrome is, in fact, the photoreceptor for all photoperiodic responses, of which a very large number had been described.

In the remainder of this chapter attention will be directed to the role of phytochrome in photoperiodism, particularly reproductive photoperi-odism, in angiosperms, and the interaction of phytochrome with endoge-nous circadian rhythms in photoperiodic timing. Discussion of evidence for flowering hormones and the effects of some of the known kinds of hormones on flowering will comprise the closing sections.

Photoperiodism

Photoperiodism is a response to the relative lengths of day and night, or more specifically, a response to the duration and timing of light and dark periods. This phenomenon is an integral part of the overall adaptation of many kinds of organisms—plant and animal—to their natural environments. Among the kinds of organisms which exhibit photoperiodism are at least some species among nonvascular and vascular cryptogams, gymnosperms, angiosperms, insects, fishes, reptiles, birds, and mammals.

In all but equatorial areas, of course, daylength changes at a constant rate with the change of the seasons. Photoperiod is therefore the "least variable variable in nature," as has often been noted. When this fact is appreciated, it does not seem at all surprising that photoperiodism has been of such prominent significance in the evolution of both plant and animal kingdoms.

Many processes and developmental events in both the vegetative and flowering stages of the life cycles of angiosperms are either controlled or greatly influenced by the duration and timing of daily light and dark periods. Among the many known examples of aspects of vegetative development which are affected by daylength, in at least some species, are: (1) onset and breaking of bud dormancy in woody perennials, (2) leaf abscission in deciduous trees and shrubs, (3) bulb and tuber formation in herbaceous species, (4) seed germination, (5) cessation of cambial activity, and (6) development of frost resistance. Such purely vegetative responses frequently are termed "vegetative photoperiodism." More familiar to most students of biology, perhaps, is the effect of daylength on the flowering behavior of many species of angiosperms. It was primarily this phenomenon of "reproductive photoperiodism" in angiosperms which was described in the classical work of W. W. Garner and H. A. Allard in 1920. Garner and Allard discovered that angiosperms could be classified in three main groups, according to the ways in which flowering is affected by daylength. These groups are the short-day plants, long-day plants, and day-neutral plants. The short-day plants were said to be those which flower in response to daylengths shorter than some maximum. Long-day plants were identified as those that flower in response to daylengths longer than some critical duration. Day-neutral plants were regarded as those in which flowering is not specifically affected by daylength.

Actually, and as might have been expected, the diversity of response types is much greater than is reflected in the simple classification initially devised by Garner and Allard. For example, a few species flower only on intermediate daylength, that is, when the days are neither too short nor too long. Other species are known which flower in response to a period of short days followed by a period of long days (short–long-day plants), and a few others flower in response to long days followed by short days (long–short-day plants). Winter annuals (e.g., winter varieties of wheat and rye) and biennials (e.g., carrot, cabbage, biennial henbane) require vernalization followed by long days to flower. "Vernalization" is an induction or acceleration of flowering by low temperature. Winter annuals normally are vernalized as seedlings, and biennials after the first season of growth.

Complicating the situation further, there are, in the categories of both short-day plants and long-day plants, species and varieties which exhibit more-or-less obligatory or qualitative responses and others which exhibit only facultative or quantitative responses. An obligate short-day plant typically has quite exacting requirements for photoperiodic floral induction and will not flower in their absence, whereas a facultative short-day plant does not exhibit an all-or-none response but flowers more readily or more vigorously if it experiences relatively short daylength.

In the preceding discussion emphasis has been placed upon daylength. This is somewhat misleading, however, in that photoperiodism is, as stated earlier, a response to the duration and timing of light and dark periods. The single most important controlling factor is the length of the uninterrupted dark period. For all photoperiodic responses occurring in natural 24-hour light–dark cycles, it is possible to speak of both "critical daylength" and "critical nightlength." Critical daylength, in the case of a long-day plant, is the minimum daylength required for flowering; in the case of a short-day plant, it is the maximum daylength on which flowering can occur. The critical nightlength for a short-day plant is the minimum period of darkness required for flowering; for a long-day plant it is the maximum period of darkness on which flowering can occur. Thus for short-day plants the daylength must be shorter than the critical daylength and the dark period longer than the critical nightlength if flowering is to occur. For long-day plants, on the contrary, the daylength must be longer than the critical daylength and the dark period shorter than the critical nightlength for flowering to occur. The distinction between short-day plants and long-day plants has nothing to do with the absolute values of the critical daylength; the distinction is whether flowering is promoted by daylengths shorter or longer than the critical daylength. For example, the critical daylength for *Xanthium strumarium,* an absolute short-day plant, is about 15.5 hours, and for *Hyoscyamus niger,* an obligate long-day plant, about 11 hours. Both species flower well in response to 12- to 14-hour photoperiods in 24-hour light–dark cycles.

A general summary description of the requirements for photoperiodic induction of flowering in obligate short-day plants and long-day plants can be stated as follows:

> *Short-day plants* require exposure to a certain minimum number of photoinductive cycles (which must be consecutive for maximum effectiveness) in each of which an uninterrupted dark period equals or exceeds a critical duration, after attaining "ripeness to flower."
>
> *Long-day plants* require exposure to a certain minimum number of cycles (not necessarily consecutive) in each of which the photoperiod equals or exceeds some critical duration, after attaining "ripeness to flower;" they do not require a dark period.

Examples of obligate short-day plants are cocklebur (*Xanthium strumarium*), Japanese morning glory (*Pharbitis nil*), kalanchoë (*Kalanchoë blossfeldiana*), poinsettia (*Euphorbia pulcherrima*), Maryland Mammoth tobacco (*Nicotiana tabacum*), and the common weed *Chenopodium rubrum*. Among known obligate long-day plants are black henbane (*Hyoscyamus niger*), Darnel ryegrass (*Lolium temulentum*), sedum (*Sedum spectabile*), spinach (*Spinacia oleracea*), and some varieties of barley (*Hordeum vulgare*). Tomato (*Lycopersicum esculentum*) and corn (*Zea mays*) are familiar day-neutral plants.

The age or amount of development which a plant must attain before it is "ripe to flower" varies markedly with species. Some plants, such as *Pharbitis nil* and *Chenopodium rubrum*, can be induced to flower in the seedling stage. Others, such as *Xanthium strumarium* and *Hyoscyamus niger*, must be a few weeks of age, and some trees must be several years old before they can be induced to flower.

Species also differ markedly with respect to the number of photoinductive cycles required for flowering. Some very sensitive species, such as *Xanthium strumarium*, *Lolium temulentum*, and *Pharbitis nil*, will respond to a single photoinductive cycle. Others require several cycles and, for maximum effectiveness, in the case of short-day plants, the cycles must be consecutive. In those species which respond to a single photoinductive cycle, as in other species, the response (rate and vigor) increases as the number of cycles increases, up to some maximum number.

Previously reference was made to the night interruption phenomenon, which is manifested as a nullification of the promoting effect of an otherwise inductive dark period on the flowering of short-day plants and nullification of the inhibitory effect of a long dark period for long-day plants. Consider, for examples, certain hypothetical photoperiodic treatments of *Xanthium strumarium* (an obligate short-day plant with a critical daylength of 15.5 hours and critical nightlength, therefore, of 8.5 hours) and annual *Hyoscyamus niger* (an obligate long-day plant with a critical daylength of about 11 hours) (Table 7-2). A photoperiodic cycle consisting of a 16-hour day and an 8-hour night would not stimulate flowering in *X. strumarium* (because the dark period is too short) but would stimulate *H. niger* (because the critical daylength is equaled or exceeded). Conversely, a cycle consisting of an 8-hour day and a 16-hour night would

Table 7-2. Expected Flowering Response of *Xanthium strumarium* (SDP) and *Hyoscyamus niger* (LDP) to Various Photoperiodic Treatments

Photoperiodic treatment	Expected flowering response	
	X. strumarium	*H. niger*
(1) 16-hour day, 8-hour night	−	+
(2) 8-hour day, 16-hour night	+	−
(3) 12-hour day, 12-hour night	+	+
(4) 10-hour day, 14-hour night— interrupted by light	−	+

stimulate flowering of *X. strumarium* (because the dark period exceeds the critical night) but would not evoke flowering of *H. niger* (because the dark period is too long). A photoperiodic cycle made up of a 12-hour day and a 12-hour night would stimulate flowering in both, chiefly because, for *X. strumarium,* the dark period is longer than the critical nightlength of 8.5 hours, and, for *H. niger,* the photoperiod equals or exceeds the critical daylength. A 10-hour photoperiod and a 14-hour night, interrupted at the 7th hour with a flash of high-intensity white light, would not stimulate *X. strumarium* (because the interrupted dark periods would be too brief) but stimulate *H. niger* (because the interruption of the dark period would nullify the otherwise inhibitory effect of a 14-hour night).

Determination of the action spectrum for the night interruption phenomenon would reveal that red light (in the region of 660 nm wavelength) is most effective, and that the effect of red light can be counteracted with immediate subsequent irradiation with far-red light (around 730 nm wavelength). All the criteria essential to prove that phytochrome is the photoreceptor for the night interruption phenomenon would be fully met.

As the transformation reactions for phytochrome were discovered, there quickly developed the hypothesis that "pigment shift" represents the basic timing mechanism in photoperiodic timing. "Pigment shift" is the process whereby, during darkness, all the phytochrome comes to be present as P_r. Component reactions of "pigment shift" in all plants are destruction of P_{fr} and synthesis of new phytochrome as P_r. In some dicots, a third reaction is reversion of P_{fr} to P_r in darkness. By this hypothesis the critical nightlength would simply be the time required for all the phytochrome to come to be present as P_r in darkness. However, it turned out that this hypothesis is inadequate by itself. In the few cases where "pigment shift" has been measured the process appears to be complete in only 2–4 hours or less. Moreover, it has become clear that endogenous circadian rhythms are very much involved in photoperiodic timing.

Endogenous Circadian Rhythms

A fundamentally important question regarding photoperiodism is: How do plants measure time in darkness? We have seen that the process of "pigment shift" involving phytochrome cannot, by itself, account for measurement of dark periods. What is involved evidently is a complex interaction of the phytochrome system with endogenous circadian rhythms.

Endogenous rhythms are characteristic of all living things except perhaps for bacteria. An endogenous rhythm was discussed in Chapter 1 in relation to bud dormancy in a tropical tree. In that example the rhythm had a period of approximately 26 days. It is to be remembered that there are many endogenous rhythms known for plants, and there is variation among all the known rhythms with respect to period length. However, the most familiar type of rhythms are the endogenous "circadian" rhythms or

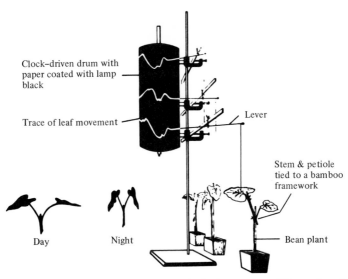

Clock–driven drum with
paper coated with lamp
black

Trace of leaf movement

Lever

Stem & petiole
tied to a bamboo
framework

Day   Night

Bean plant

Figure 7-14. Circadian movements of a leaf of a bean plant (*Phaseolus vitellinus*) show a rising in the day and a falling at night. (From *Plant Physiology,* 2nd ed., by Frank B. Salisbury and Cleon W. Ross. © 1978 by Wadsworth Publishing Company, Inc., Belmont, California 94002. Reprinted by permission of the publisher.)

"biological clocks" which have a period of approximately 24 hours ("circa" = about; "diem" = day) (Fig. 7-14; cf. Figs. 1-11 and 1-13).

Under normal circumstances, of course, plants grow in an environment in which a number of parameters (e.g., light intensity and temperature) fluctuate rhythmically with a period of 24 hours. It is not surprising, therefore, that plants have adapted their physiological activity to cycles of approximately 24 hours. Rhythmic behavior in plants has been known for nearly 2000 years, yet the mechanisms underlying endogenous rhythms, circadian and otherwise, remain largely a mystery. This is extremely challenging, because of the very great biological significance of "biological clocks." Presently we understand only some of the more salient features of endogenous circadian rhythms such as: (1) the natural free-running period is not precisely 24 hours but usually is somewhere between 21 and 28 hours; (2) the rhythms are not merely the consequence of fluctuations in physical environmental conditions, although it often is possible to entrain circadian rhythms to periods of other than about 24 hours by manipulation of physical factors; (3) however, even an entrained system normally returns to its natural value immediately if the organism is returned to the natural environment; and (4) endogenous circadian rhythms are only slightly affected by temperature over a wide range.

Experiments conducted with photoperiodic flowering responses have indicated rather vividly the role of endogenous circadian rhythms in photoperiodic reactions, as Bünning suggested is the case in the early 1930s.

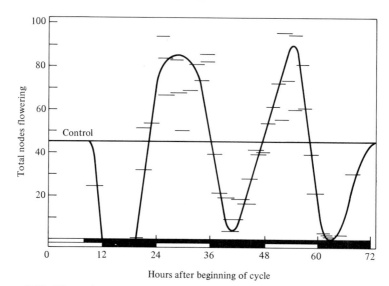

Hours after beginning of cycle

Figure 7-15. Flowering response of Biloxi soybean (a short-day plant) to 4-hour light interruptions given at various times during a 72-hour cycle. Lines indicate the 4-hour interruption periods. The control plants were exposed to seven cycles, each cycle consisting of an 8-hour photoperiod followed by a 64-hour dark period. For the various treatments the 64-hour dark period was interrupted at various times by illumination with high-intensity light of 4 hours duration. Flowering of the controls with no light interruptions for 72-hour-long cycles is represented by the horizontal line across the graph. The upper bar on the bottom of the graph indicates the 8 hours of light and the 64 hours of darkness given in each of the seven cycles to which the plants were exposed. The lower bar indicates the daily cycle of light and darkness under normal conditions. (Redrawn, with permission, from: Hamner, K. 1963. Endogenous rhythms in controlled environments. In: L. T. Evans, ed. *Environmental Control of Plant Growth.* Pp. 215–232. Copyright by Academic Press, Inc., New York.)

A classical experiment with the Biloxi variety of soybean (*Glycine max*) (Fig. 7-15) is illustrative of this very crucial point. K. C. Hamner exposed Biloxi soybeans to seven cycles of 8 hours of light followed by 64 hours of darkness and interrupted the long dark period with light at various times. The plants had been previously grown ("entrained") to 12-hour photo-periods alternating with 12-hour dark periods. The striking result was that the night interruption phenomenon (manifested as inhibition of flowering) showed three peaks, approximately 24 hours apart and corresponding to the times when the plants normally would have been in darkness. Obviously, from all that is known about the process, "pigment shift" was complete during the early hours of the 64-hour dark period. Moreover, we can confidently assume that night interruption at any time caused a certain amount of P_r to be converted to P_{fr}. Yet the consequence of P_{fr} presence depended clearly upon when interruption of the dark period occurred. Or, to put it another way, the susceptibility of the flowering process, specifically the dark processes leading to flowering, to inhibition by P_{fr} varied with the timing of the interruption of the dark period.

In summary, it appears that endogenous circadian rhythms play the basic and fundamental role in timing of light and dark periods in angiosperms generally. In photoperiodic responses, the phytochrome system seems to be superimposed as a modifying influence upon the basic timing mechanism. The nature of the control of timing in photoperiodic phenomena seems to reside in a complex interaction between the state of the phytochrome and the endogenous circadian rhythm.

One important final point remains to be clarified regarding the role of phytochrome in the flowering of photoperiodically sensitive species. One might be led to conclude, erroneously, that P_{fr} is only inhibitory to flowering of short-day plants but that it is essential to flowering of long-day plants. As a matter of fact, the evidence is clear that P_{fr} is essential to the flowering of both. The difference evidently is only a quantitative one. Short-day plants depend upon P_{fr} action occurring during a photoperiod which must not be too long (longer than the critical daylength), and they must be free of the presence of P_{fr} during a certain period of darkness. Long-day plants, in contrast, either have a greater need for P_{fr} action or are less sensitive to P_{fr} action, since they flower best when P_{fr} is continually present (i.e., in constant light).

Hormonal Regulation of Flowering

Within less than two decades after the discovery of photoperiodism, the hypothesis developed that one or more specific flowering hormones are responsible for floral initiation. Chailakhyan (1936) coined the word "florigen" for the hypothetical flowering hormone. A short time later, Melchers (1939) suggested the term "vernalin" for the hypothetical stimulus thought to develop during vernalization of cold-requiring plants. Many years later Chailakhyan proposed the term "anthesins" for hypothetical flowering hormones (Chailakhyan, 1970) and gave emphasis to the role GAs play in flowering of many plants.

Early experiments, conducted in the 1930s and 1940s, with classical obligate short-day plants such as cocklebur (*Xanthium strumarium*), which responds to a single photoinductive cycle (Fig. 7-16), yielded results which provided a very logical basis for the florigen (anthesin) concept. It was observed that, in the case of cocklebur but not all short-day plants, induction of a single leaf was sufficient to cause the plant to flower systemically (Fig. 7-17). Furthermore, a single photoinduced leaf or whole plant could be grafted onto another plant kept on noninductive long days and the latter would flower (Fig. 7-18). These and countless related experiments involving a large number of photoperiodically sensitive species suggested strongly that the leaf is the organ which perceives the pho-

Cocklebur grown
on long days

One cycle of short
day and long night

Return to
long day

Flowers
after
6–14 days

Figure 7-16. Photoperiodic induction in the cocklebur by a single short-day cycle. (Redrawn, with permission, from Bonner and Galston, 1952.)

toperiodic stimulus, that phytochrome is the photoreceptor, and that a substance or substances is formed in the leaves of short-day plants which is translocated to vegetative meristems and causes their morphogenetic transformation into floral meristems. Furthermore, from numerous grafting experiments, it appeared that there was only one "florigen," or if two or more substances, they were physiologically equivalent among many

Long days

This leaf
subjected to
short days.
Entire plant
flowers

Cocklebur remains
vegetative on long
days

Cocklebur flowers even
on long days if one leaf
is given short days

Figure 7-17. Photoinduction of a single leaf causes cocklebur to flower systemically. (Redrawn, with permission, from Bonner and Galston, 1952.)

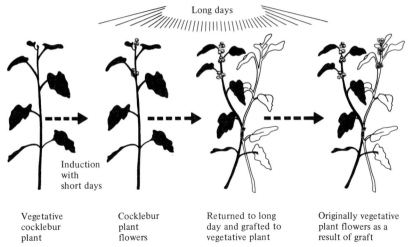

Long days

Induction
with
short days

Vegetative	Cocklebur	Returned to long	Originally vegetative
cocklebur	plant	day and grafted to	plant flowers as a
plant	flowers	vegetative plant	result of graft

Figure 7-18. The flowering stimulus can be transmitted from plant to plant across a graft union. (Redrawn, with permission, from Bonner and Galston, 1952.)

species. The various photoperiodic response types would differ not in their requirement for florigen but in the environmental requirements for florigen production. For example, "ripe-to-flower" vegetative scions of *Sedum spectabile* (a long-day plant) could be grafted onto stocks of *Kalanchoë blossfeldiana* (a short-day plant) and the grafted plants maintained under short-day conditions and the *Sedum spectabile* would flower (Fig. 7-19). The reciprocal graft under long days worked equally well. A logical explanation of these results is that a florigen was formed in the leaves of the stock under photoperiodic conditions inductive for it which passed the graft union and caused flowering of the scions under noninductive conditions. By 1963, approximately 25 such graft unions had proved successful. The 25 cases involved species from long-day, short-day, and day-neutral types.

With such evidence compatible with a florigen concept, a logical next step would be to attempt to isolate and chemically characterize the flowering stimulus or stimuli. In fact, there have been many such efforts. Generally the attempts have ended in failure; occasionally there have been modest successes. One group of investigations, which looked particularly promising, was conducted by Richard G. Lincoln and associates at Long Beach State College (1961, 1966) and by K. C. Hamner and associates (1970) at the University of California at Los Angeles. In this work, extracts were prepared from the leaves and buds of flowering cocklebur (*Xanthium strumarium*) which stimulated flowering in the short-day plant *Lemna perpusilla* under noninductive conditions and had slight flower-inducing activity also toward vegetative cocklebur plants maintained under noninductive conditions if supplemented with GA (Tables 7-3 and 7-4). "Florigenic acid" is the name given to the yet unidentified active sub-

Table 7-3. Effects of an Extract from *Xanthium strumarium* on Flowering of Cockelbur[a]

Treatment, extract or chemical	Flowering stage				
	0	1	2–5	6–10	>10
Flowering plants	91	9	0	0	0
Flowering plants + GA	11	13	44	21	11
Vegetative plants	97	3	0	0	0
Vegetative plants + GA	84	16	0	0	0
Gibberellic acid	88	12	0	0	0
Water	100	0	0	0	0

[a] Extract activity evaluated by the *Xanthium* bioassay. One hundred plants were used in each of the six treatments. Ten plants were used to evaluate each of the 20 extracts from flowering or vegetative plants. The figures represent the total number of plants in each stage of flowering. Plants in stage 2 or above are considered to be flowering. GA, gibberellic acid. (With permission from Hodson, H. K. and K. C. Hamner. 1970. Floral inducing extract from *Xanthium. Science* **167**: 384–385, 23 January 1970. Copyright 1970 by the American Association for the Advancement of Science.)

stance or substances in the extracts. Active extracts have been prepared from day-neutral sunflower plants and from a fungus (*Calonectria*) but not from vegetative cocklebur plants.

Thus, as Frank B. Salisbury and Cleon W. Ross (1978) have noted, "There is much circumstantial evidence that flower initiation is controlled by hormones: one or more positive acting florigens and one or more negatively acting inhibitors. These substances remain to be identified." That certain of the known hormones are involved in regulation of flowering is

Table 7-4. Effects of an Extract from *Xanthium strumarium* on Flowering of *Lemna purpusilla*[a]

Treatment	Plants flowering (%)									
Flowering plants	45	31	54	37	58	46	45	51	39	60
Flowering plants + GA	0	0	0	0	0	0	0	0	0	0
Vegetative plants	0	0	9	0	7	0	0	0	2	0
Vegetative plants + GA	0	0	0	0	0	0	0	0	0	0
Gibberellic acid	0	0	0	0	0	0	0	0	0	0
Water	0	0	5	0	11	3	0	0	1	5

[a] Extract activity evaluated by the *Lemna* bioassay. A total of 30,000 plants were used to evaluate the activity of the 20 extracts. The extracts were tested with and without the addition of gibberellic acid (GA). Each figure represents the average response of five lots of 100 plants each. (With permission from Hodson, H. K. and K. C. Hamner. 1970. Floral inducing extract from *Xanthium. Science* **167**: 384–385. 23 January 1970. Copyright 1970 by the American Association for the Advancement of Science.)

Figure 7-19. Transmission of flowering stimulus from the short-day plant *Kalanchoë blossfeldiana* (lower part) to the long-day plant *Sedum spectabile* (upper part) under short-day conditions. The axillary shoots of *Kalanchoöe* flowered under the influence of short days. The *Sedum* scion was induced to flower also under short-day conditions by the *Kalanchoë* stock. The photograph was taken 97 days after grafting. (By permission, from Zeevaart, 1958; see also Zeevaart, 1962.)

definite as indicated by numerous investigations conducted with GAs since circa 1957 and with ABA since circa 1967.

Anton Lang first reported in 1957 that exogenous GA could cause flowering in numerous species of long-day and vernalization-requiring plants under noninductive environmental conditions (Fig. 7-20), but GA is not the long sought after florigen. One reason is that GAs do not cause flowering of short-day plants under noninductive conditions, or even in all long-day plants. Moreover, there is now good evidence that the flowering and flower-bearing stem elongation (bolting) are separate processes in plants such as *Silene armeria* (Cleland and Zeevaart, 1970), with GA promoting stem elongation only. The latter conclusion was based on the findings that: (1) treatment of *Silene armeria* with Amo-1618, an inhibitor of GA biosynthesis, inhibited stem elongation but not flowering under long-day conditions; and (2) with GA treatment on short days, stem elongation sometimes occurred in the absence of flowering. Thus it seems clear that although flowering and stem elongation are closely related in rosette-type long-day plants under normal conditions, the two processes can be experimentally separated and thus represent separate developmen-

Figure 7-20. Carrot (*Daucus carota*) plants showing the application of GA as a substitute for a cold requirement for flowering. Treatments: Left, no GA, no cold treatment. Center, 10 μg GA per day, no cold treatment. Right, no GA but 8 weeks cold treatment. The plants were grown in long days. (From Lang, 1957, with permission; photo courtesy of A. Lang, 1978.)

tal processes. While GA is not "florigen," it conceivably might be "vernalin."

During the early investigations of the effects of exogenous ABA on plants, it was reported that this growth-inhibiting hormone could cause flowering of certain short-day plants (e.g., *Chenopodium rubrum, Pharbitis nil,* and some others) under long-day conditions. But ABA not a florigen because there is no effect of exogenous ABA on flowering of some short-day plants or on long-day plants. It may be one of the postulated flowering inhibitors, which have long been thought to be produced in the leaves of long-day plants subjected to noninductive short-day conditions.

Thus, as the final section of the last chapter of this discussion of plant hormones ends, it is clear that much remains to be discovered about the hormonal regulation of flowering as well as many other aspects of hormonal regulation of plant growth and development. Undoubtedly, fascinating new discoveries lie ahead.

References

Attridge, T. H. and C. B. Johnson. 1976. Photocontrol of enyzme levels. In: H. Smith, ed. *Light and Plant Development.* Butterworth & Company, Limited, London. Pp. 185–192.

Bonner, J. and A. W. Galston. 1952. *Principles of Plant Physiology.* W. H. Freeman and Company, San Francisco.

Borthwick, H. 1972. History of phytochrome. In: K. Mitrakos and W. Shropshire, Jr., eds. *Phytochrome.* Academic Press, New York. Pp. 3–23.

Briggs, W. R. 1976. The nature of the blue light photoreceptor in higher plants and fungi. In: H. Smith, ed. *Light and Plant Development.* Butterworth & Company, Limited, London. Pp. 7–18.

Briggs, W. R. and H. V. Rice. 1972. Phytochrome: chemical and physical properties and mechanism of action. *Ann. Rev. Plant Physiol.* **23**: 293–334.

Briggs, W. R. and H. W. Siegelman. 1965. Distribution of phytochrome in etiolated seedlings. *Plant Physiol.* **40**: 934–941.

Browning, G. and P. F. Saunders. 1977. Membrane localised gibberellins A_9 and A_4 in wheat chloroplasts. *Nature (London)* **265**: 375–377.

Bünning, E. 1973. *The Physiological Clock.* 3rd ed. Academic Press, New York.

Bünning, E. 1977. Fifty years of research in the wake of Wilheim Pfeffer. *Ann. Rev. Plant Physiol.* **28**: 1–22.

Butler, W. L., K. H. Norris, H. W. Siegelman, and S. B. Hendricks. 1959. Detection, assay, and preliminary purification of the pigment controlling photoresponsive development of plants. *Proc. Natl. Acad. Sci. U.S.A.* **45**: 1703–1708.

Chailakhyan, M. Kh. 1968. Internal factors of plant flowering. *Ann. Rev. Plant Physiol.* **19**: 1–36.

Chailakhyan, M. Kh. 1970. Flowering and photoperiodism of plants. *Plant Sci. Bull.* **3**(3): 1–7.

Chailakhyan, M. Kh. 1975. Forty years of research on the hormonal basis of plant development—some personal reflections. *Bot. Rev.* **41**: 1–29.

Clarkson, D. T. and W. S. Hillman. 1967. Apparent phytochrome synthesis in *Pisum* tissue. *Nature (London)* **213**: 468–470.

Clarkson, D. T. and W. S. Hillman. 1968. Stable concentrations of phytochrome in Pisum under continuous illumination with red light. *Plant Physiol.* **43**: 88–92.

Cleland, C. F. and J. A. D. Zeevaart. 1970. Gibberellins in relation to flowering and stem elongation in the long day plant *Silene armeria*. *Plant Physiol.* **46**: 392–400.

Cooke, R. J. and R. E. Kendrick. 1976. Phytochrome controlled gibberellin metabolism in etioplast envelopes. *Planta* **131**: 303–307.

Cooke, R. J. and P. F. Saunders. 1975. Phytochrome mediated changes in extractable gibberellin activity in a cell-free system from etiolated wheat leaves. *Planta* **123**: 299–302.

Cumming, B. G. and E. Wagner. 1968. Rhythmic processes in plants. *Ann. Rev. Plant Physiol.* **19**: 381–416.

Downs, R. J. and H. Hellmers. 1975. *Environment and the Experimental Control of Plant Growth*. Academic Press, New York.

Ellis, R. J. and M. R. Hartley. 1971. Sites of synthesis of chloroplast proteins. *Nature (London) New Biol.* **233**: 193–196.

Evans, A. and H. Smith. 1976a. Spectrophotometric evidence for the presence of phytochrome in the envelope membranes of barley etioplasts. *Nature (London)* **259**: 323–325.

Evans, A. and H. Smith. 1976b. Localization of phytochrome in etioplasts and its regulation *in vitro* of gibberellin levels. *Proc. Natl. Acad. Sci. U.S.A.* **73**: 138–142.

Evans, L. T., 1975. *Daylength and the Flowering of Plants*. W. A. Benjamin, Inc., Menlo Park, California.

Flint, L. H. and E. D. McAlister. 1935. Wave lengths of radiation in the visible spectrum inhibiting the germination of light-sensitive lettuce seed. *Smithsonian Inst. Misc. Coll.* **94**(5): 1–11.

Flint, L. H. and E. D. McAlister. 1937. Wavelengths of radiation in the visible spectrum promoting the germination of light-sensitive lettuce seed. *Smithsonian Inst. Misc. Coll.* **96**: 1–8.

Fondeville, J. C., H. A. Borthwick, and S. B. Hendricks. 1966. Leaflet movement of *Mimosa pudica* L. indicative of phytochrome action. *Planta* **69**: 357–364.

Galston, A. W. 1974. Plant photobiology in the last half-century. *Plant Physiol.* **54**: 427–436.

Galston, A. W. and R. L. Satter. 1976. Light, clocks, and ion flux: an analysis of leaf movement. In: H. Smith, ed. *Light and Plant Development*. Butterworth & Company, Limited, London. Pp. 159–184.

Garner, W. W. and H. A. Allard. 1920. Effect of the relative length of day and night and other factors of the environment on growth and reproduction in plants. *J. Agr. Res.* **18**: 553–606.

Garner, W. W. and H. A. Allard. 1923. Further studies in photoperiodism, the response of the plant to the relative length of day and night. *J. Agr. Res.* **23**: 871–920.

Hamner, K. 1963. Endogenous rhythms in controlled environments. In: L. T. Evans, ed. *Environmental Control of Plant Growth*. Academic Press, New York. Pp. 215–232.

Hampp, R. and H. W. Schmidt. 1977. Regulation of membrane properties of mitochondria and plastids during chloroplast development. I. The action of phytochrome in situ. *Z. Pflanzenphysiol.* **82**: 68–77.

Haupt, W. 1963. Photoreceptorprobleme der Chloroplastenbewegung. *Ber. Deutsch. Bot. Gesellschaft* **76**: 313–322.

Haupt, W. 1973. Role of light in chloroplast movement. *BioScience* **23**: 289–296.

Hillman, W. S. 1962. *The Physiology of Flowering.* Holt, Rinehart & Winston, New York.

Hillman, W. S. 1976. Biological rhythms and physiological timing. *Ann. Rev. Plant Physiol.* **27**: 159–179.

Hodson, H. K. and K. C. Hamner. 1970. Floral inducing extract from *Xanthium. Science* **167**: 384–385.

Jaffe, M. J. 1968. Phytochrome-mediated bioelectric potentials in mung bean seedlings. *Science* **162**: 1016–1017.

Jaffe, M. J. and A. W. Galston. 1967. Phytochrome control of rapid nyctinastic movements and membrane permeability in *Albizzia julibrissin. Planta* **77**: 135–141.

Lang, A. 1957. The effect of gibberellin upon flower formation. *Proc. Natl. Acad. Sci. U.S.A.* **43**: 709–717.

Lhoste, J. M. 1972. Some physical aspects of the phytochrome phototransformation. In: K. Mitrakos and W. Shropshire, Jr., eds. *Phytochrome.* Academic Press, New York. Pp. 47–74.

Lincoln, R. G., A. Cunningham, B. H. Carpenter, J. Alexander, and D. L. Mayfield. 1966. Florigenic acid from fungal culture. *Plant Physiol.* **41**: 1079–1080.

Lincoln, R. G., D. L. Mayfield, and A. Cunningham. 1961. Preparation of a floral initiating extract from Xanthium. *Science* **133**: 756.

Marmé, D. 1977. Phytochrome: membranes as possible sites of primary action. *Ann. Rev. Plant Physiol.* **27**: 173–198.

Mitrakos, K. and W. Shropshire, Jr., eds. 1972. *Phytochrome.* Academic Press, New York.

Mohr, H., C. Huault, H. Lange, L. Lohmann, I. Rissland, and M. Weidner. 1968. Lag-phases in phytochrome-mediated enzyme synthesis (PAL). *Planta* **83**: 267–275.

Pharis, R. P. and C. G. Kuo. 1977. Physiology of gibberellins in conifers. *Can. J. Forest. Res.* **7**: 299–325.

Pratt, L. H. 1976. Re-examination of photochemical properties and absorption characteristics of phytochrome using high-molecular-weight preparations. In: H. Smith, ed. *Light and Plant Development.* Butterworth & Company, Limited, London. Pp. 19–30.

Quail, P. H. 1976. Phytochrome. In: Bonner, J. and J. E. Varner, eds. *Plant Biochemistry.* 3rd ed. Academic Press, New York. Pp. 683–711.

Rüdiger, W. 1972. Chemistry of phytochrome chromophore. In: Mitrakos, K. and W. Shropshire, Jr., eds. *Phytochrome.* Academic Press, New York. Pp. 129–141.

Salisbury, F. B. 1963. *The Flowering Process.* Pergamon Press, New York.

Salisbury, F. B. 1965. The initiation of flowering. *Endeavour* **24**: 74–80.

Salisbury, F. B. 1971. *The Biology of Flowering.* Natural History Press, Garden City, New York.

Salisbury, F. B. and C. W. Ross. 1978. *Plant Physiology.* 2nd ed. Wadsworth Publishing Company, Belmont, California.

Satter, R. L. and A. W. Galston. 1976. The physiological functions of phytochrome. In: L. W. Goodwin, ed. *Chemistry and Biochemistry of Plant Pigments.* Vol. 1. Academic Press, New York. Pp. 680–735.

Satter, R. L., P. Marinoff, and A. W. Galston. 1972. Phytochrome-controlled nyctinasty in *Albizzia julibrissin.* IV. Auxin effects on leaflet movement and K flux. *Plant Physiol.* **50**: 235–241.

Schmidt, H. W. and R. Hampp. 1977. Regulation of membrane properties of mitochondria and plastids during chloroplast development. II. The action of phytochrome in a cell-free system. *Z. Pflanzenphysiol.* **82**: 428–434.

Schopfer, P. 1977. Phytochrome control of enzymes. *Ann. Rev. Plant Physiol.* **28**: 223–252.

Siegelman, H. W. and W. L. Butler. 1965. Properties of phytochrome. *Ann. Rev. Plant Physiol.* **16**: 383–392.

Siegelman, H. W. and E. M. Firer. 1964. Purification of phytochrome from oat seedlings. *Biochemistry* **3**: 418–423.

Smith, H., ed. 1976. *Light and Plant Development.* Butterworth & Company, Limited, London.

Stoddart, J. L. 1968. The association of gibberellin-like activity with the chloroplast fraction of leaf homogenates. *Planta* **81**: 106–112.

Stoddart, J. L. 1969. Incorporation of kaurenoic acid into gibberellins by chloroplast preparations of *Brassica oleracea. Phytochemistry* **8**: 831–837.

Sweeney, B. M. 1969. *Rhythmic Phenomena in Plants.* Academic Press, New York.

Tanada, T. 1968. A rapid photoreversible response of barley root tips in the presence of 3-indoleacetic acid. *Proc. Natl. Acad. Sci. U.S.A.* **59**: 376–379.

Travis, R. L., J. L. Key, and C. W. Ross. 1974. Activation of 80S maize ribosomes by red light treatment of dark-grown seedlings. *Plant Physiol.* **53**: 28–31.

Vince-Prue, D. 1975. *Photoperiodism in Plants.* McGraw-Hill Book Company, New York.

Wellburn, A. R. and R. Hampp. 1976. Fluxes of gibberellic and abscisic acids, together with that of adenosine 3', 5'-cyclic phosphate, across plastid envelopes during development. *Planta* **131**: 95–96.

Zeevaart, J. A. D. 1958. Flower formation as studied by grafting. *Mededel. Landbouwhogeschool Wageningen* **58**: 1–88.

Zeevaart, J. A. D. 1962. Physiology of flowering. *Science* **137**: 723–731.

Zeevaart, J. A. D. 1976. Physiology of flower formation. *Ann. Rev. Plant Physiol.* **27**: 321–348.

Index